THE MICROPHONE BOOK

Second edition

2

THE MICROPHONE BOOK

Second edition

John Eargle

ELSEVIER

AMSTERDAM • BOSTON • HEIDELBERG • LONDON • NEW YORK • OXFORD
PARIS • SAN DIEGO • SAN FRANCISCO • SINGAPORE • SYDNEY • TOKYO
Focal Press is an imprint of Elsevier

Focal
Press

Focal Press
An Imprint of Elsevier
Linacre House, Jordan Hill, Oxford OX2 8DP
30 Corporate Drive, Burlington MA 01803

First published 2004

British Library Cataloguing in Publication Data
A catalogue record for this book is available from the British Library

Library of Congress Cataloguing in Publication Data
A catalogue record for this book is available from the Library of Congress

ISBN-13: 978-0-240-51961-6
ISBN-10:   0-240-51961-2

For information on all Focal Press publications visit our website at:
www.focalpress.com

Typeset by Newgen Imaging Systems (P) Ltd, Chennai, India
Printed and bound in Great Britain

CONTENTS

PREFACE TO THE FIRST EDITION

Most sound engineers will agree that the microphone is the most important element in any audio chain, and certainly the dazzling array of current models, including many that are half-a-century old, attests to that fact. My love affair with the microphone began when I was in my teens and got my hands on a home-type disc recorder. Its crystal microphone was primitive, but I was nonetheless hooked. The sound bug had bitten me, and, years of music schooling not withstanding, it was inevitable that I would one day end up a recording engineer.

About thirty years ago I began teaching recording technology at various summer educational programs, notably those at the Eastman School of Music and later at the Aspen Music Festival and Peabody Conservatory. I made an effort at that time to learn the fundamentals of microphone performance and basic design parameters, and my *Microphone Handbook*, published in 1981, was a big step forward in producing a text for some of the earliest collegiate programs in recording technology. This new book from Focal Press presents the technology in greater depth and detail and, equally important, expands on contemporary usage and applications.

The Microphone Book is organized so that both advanced students in engineering and design and young people targeting a career in audio can learn from it. Chapter 1 presents a short history of the microphone. While Chapters 2 through 6 present some mathematically intensive material, their clear graphics will be understandable to those with little technical background. Chapters 7 through 10 deal with practical matters such as standards, the microphone-studio electronic interface, and all types of accessories.

Chapters 11 through 17 cover the major applications areas, with emphasis on the creative aspects of music recording in stereo and surround sound, broadcast/communications, and speech/music reinforcement. Chapter 18 presents an overview of advanced development in microphone arrays, and Chapter 19 presents helpful hints on microphone maintenance and checkout. The book ends with a comprehensive microphone bibliography and index.

I owe much to Leo Beranek's masterful 1954 *Acoustics* text, A. E. Robertson's little known work, *Microphones,* which was written for the BBC in 1951, as well as the American Institute of Physics' *Handbook of Condenser Microphones.* As always, Harry Olson's books came to my aid with their encyclopedic coverage of everything audio.

Beyond these four major sources, any writer on microphones must rely on technical journals and on-going discussions with both users and manufacturers in the field. I would like to single out for special thanks the following persons for their extended technical dialogue: Norbert Sobol (AKG Acoustics), Jörg Wuttke (Schoeps GmbH), David Josephson (Josephson Engineering), Keishi Imanaga (Sanken Microphones), and in earlier days Steve Temmer and Hugh Allen of Gotham Audio. Numerous manufacturers have given permission for the use of photographs and drawings, and they are credited with each usage in the book.

John Eargle
April 2001

PREFACE TO THE SECOND EDITION

The second edition of The Microphone Book follows the same broad subject outline as the first edition. Most of the fundamental chapters have been updated to reflect new models and electronic technology, while those chapters dealing with applications have been significantly broadened in their coverage.

The rapid growth of surround sound technology merits a new chapter of its own, dealing not only with traditional techniques but also with late developments in virtual imaging and the creation of imaging that conveys parallax in the holographic sense.

Likewise, the chapter on microphone arrays has been expanded to include discussions of adaptive systems as they involve communications and useful data reduction in music applications.

Finally, at the suggestion of many, a chapter on classic microphones has been included. Gathering information on nearly thirty models was a far more difficult task than one would ever have thought, and it was truly a labor of love.

John Eargle
Los Angeles, June 2004

SYMBOLS USED IN THIS BOOK

a	radius of diaphragm (mm); acceleration (m/s^2)
A	ampere (unit of electrical current)
AF	audio frequency
c	speed of sound (334 m/s at normal temperature)
$°C$	temperature (degrees celsius)
d	distance (m)
dB	relative level (decibel)
$dB(A)$	A-weighted sound pressure level
dBu	signal voltage level (re 0.775 volt rms)
D_C	critical distance (m)
DI	directivity index (dB)
E	voltage (volt dc)
$e(t)$	signal voltage (volt rms)
f	frequency in hertz (s^{-1})
HF	high frequency
Hz	frequency (hertz, cycles per second)
I	acoustical intensity (W/m^2)
I	dc electrical current, ampere (Q/s)
i(t)	signal current (ampere rms)
I_0	mechanical moment of inertia (kg $\times$ m^2)
j	complex algebraic operator, equal to $\sqrt{-1}$
k	wave number (2π/λ)
kg	mass, kilogram (SI base unit)
$°K$	temperature (degrees kelvin, SI base unit)
LF	low frequency
L_P	sound pressure level (dB re 20 μPa)
L_R	reverberant sound pressure level (dB re 20 μPa)
L_N	noise sound pressure level (dB re 20 μPa)
m	meter (SI base unit)
MF	mid frequency
mm	millimeter (m $\times$ 10^{-3})
μm	micrometer or micron (m $\times$ 10^{-6})

M	microphone system sensitivity, mV/Pa
M_D	capacitor microphone base diaphragm sensitivity, V/Pa
N	force, newton (kg, m/s^2)
$p; p(t)$	rms sound pressure (N/m^2)
P	power (watt)
Q	electrical charge (coulombs, SI base unit)
Q	directivity factor
r	distance from sound source (m)
R, Ω	electrical resistance (ohm)
R	room constant (m^3 or ft^3)
RE	random efficiency of microphone (also REE)
RF	radio frequency
RH	relative humidity (%)
s	second (SI base unit)
S	surface area (m^2)
T	torque (N × m)
T, t	time (s)
T	magnetic flux density (tesla)
T_{60}	reverberation time (seconds)
T_0	diaphragm tension (newton/meter)
$torr$	atmospheric pressure; equal to mm of mercury (mmHg), or 133.322 Pa (Note: 760 torr is equal to normal atmospheric pressure at 0°C)
$u; u(t)$	air particle rms velocity (m/s)
$U, U(t)$	air volume rms velocity (m^3/s)
$x(t)$	air particle displacement (m/s)
X	mechanical, acoustical, or electrical reactance (Ω)
V	electrical voltage (voltage or potential)
Z	mechanical, acoustical or electrical resistance (Ω)
$\bar{\alpha}$	average absorption coefficient (dimensionless)
λ	wavelength of sound in air (m)
ϕ	phase, phase shift (degrees or radians)
ρ	dependent variable in polar coordinates
ρ_0	density of air (1.18 kg/m^3)
$\rho_0 c$	specific acoustical impedance of air (415 SI rayls)
θ	angle (degrees or radians), independent variable in polar coordinates
ω	$2\pi f$ (angular frequency in radians/s)
σ_m	surface mass density (kg/m^2)

C H A P T E R   1

A SHORT HISTORY OF THE MICROPHONE

INTRODUCTION

The microphone pervades our daily lives through the sound we hear on radio, television and recordings, paging in public spaces, and of course in two-way communications via telephone. In this chapter we will touch upon some of the highlights of more than 125 years of microphone development, observing in particular how most of the first 50 of these years were without the benefits of electronic amplification. The requirements of telephony, radio broadcast, general communications, and recording are also discussed, leading to some conjecture on future requirements.

THE EARLY YEARS

As children, many of us were fascinated with strings stretched between the ends of a pair of tin cans or wax paper cups, with their ability to convey speech over a limited distance. This was a purely mechano-acoustical arrangement in which vibrations generated at one end were transmitted along the string to actuate vibrations at the other end.

In 1876, Alexander Graham Bell received US patent 174,465 on the scheme shown in Figure 1–1. Here, the mechanical string was, in a sense, replaced by a wire that conducted electrical direct current, with audio signals generated and received via a moving armature transmitter and its associated receiver. Like the mechanical version, the system was reciprocal. Transmission was possible in either direction; however, the patent also illustrates the acoustical advantage of a horn to increase the driving pressure at the sending end and a complementary inverted horn to reinforce output pressure at the ear at the receiving end. Bell's further experiments with the transmitting device resulted in the liquid transmitter,

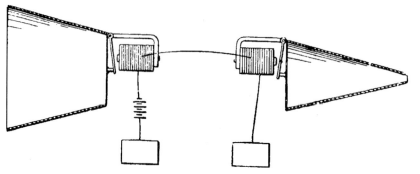

FIGURE 1-1 ————

The beginnings of
telephony; Bell's
original patent.

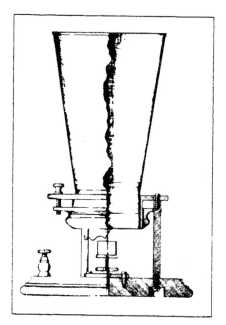

FIGURE 1-2 ————

Bell's liquid transmitter
exhibited at the
Philadelphia Centennial
Exposition of 1876.

shown in Figure 1–2, which was demonstrated at the Philadelphia
Centennial Exposition of 1876. Here, the variable contact principle
provided a more effective method of electrical signal modulation than
that afforded by the moving armature.

The variable contact principle was extended by Berliner in a patent
application in 1877 in which a steel ball was placed against a stretched
metal diaphragm, as shown in Figure 1–3. Further work in this area was
done by Blake (patents 250, 126 through 250, 129, issued in 1881), who
used a platinum bead impressed against a hard carbon disc as the variable resistance element, as shown in Figure 1–4. The measured response
of the Blake device spanned some 50 decibels over the frequency range

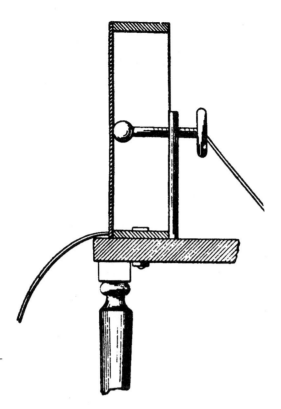

FIGURE 1-3

Berliner's variable
contact microphone.

from 380 Hz to 2000 Hz, and thus fell far short of the desired response. However, it provided a more efficient method of modulating telephone signals than earlier designs and became a standard in the Bell system for some years.

Another interim step in the development of loose contact modulation of direct current was developed in 1878 by Hughes and is shown in Figure 1–5. In this embodiment, very slight changes in the curvature of the thin wood plate diaphragm, caused by impinging sound waves, gave rise to a fairly large fluctuation in contact resistance between the carbon rod and the two mounting points. This microphone was used by Clement Ader (Scientific American, 1881) in his pioneering two-channel transmissions from the stage of the Paris Opera to a neighboring space. It was Hughes, incidentally, who first used the term *microphone*, as applied to electroacoustical devices.

The ultimate solution to telephone transmitters came with the development of loose carbon granule elements as typified by Blake's transmitter of 1888, shown in Figure 1–6. Along with the moving armature receiver, the loose carbon granule transmitter, or microphone, has dominated telephony up to the present. Quite a testimony to the inventiveness and resourcefulness of engineers working nearly 130 years ago.

THE RISE OF BROADCASTING

The carbon granule transmitter and moving armature receiver complemented each other nicely. The limitations in bandwidth and dynamic range have never been a problem in telephony, and the rather high distortion generated in the combined systems actually improved speech intelligibility by emphasizing higher frequencies. Even after the invention of electronic amplification (the de Forest audion vacuum tube, 1907), these earlier devices remained in favor.

When commercial broadcasting began in the early 1920s, there was a clear requirement for better microphones as well as loudspeakers. Western Electric, the manufacturing branch of the Bell Telephone system, was quick to respond to these needs, developing both *electrostatic* (capacitor) microphones as well as *electrodynamic* (moving conductor) microphones. The capacitor, or condenser, microphone used a fixed electrical charge on the plates of a capacitor, one of which was a moving diaphragm and the other a fixed back plate. Sound waves caused a slight variation in capacitance, which in turn was translated into a variation in the voltage across the plates. An early Western Electric capacitor microphone, developed by Wente in 1917, is shown in Figure 1–7.

While first employed as a driving element for loudspeakers and headphones, the moving coil and its close relative, the ribbon, eventually found their place in microphone design during the mid-twenties. Moving coil and ribbon microphones operate on the same principle; the electrical conducting element is place in a transverse magnetic field, and its motion generated by sound vibrations induces a voltage across the conducting element. Under the direction of Harry Olson, Radio Corporation of America (RCA) was responsible for development and beneficial exploitation of the ribbon microphone during the 1930s and 1940s.

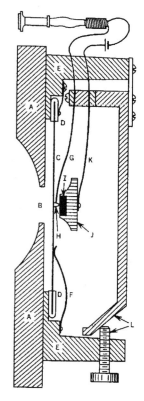

FIGURE 1-4

Blake's carbon disc microphone.

FIGURE 1-5

Hughes' carbon rod microphone.

THE RISE OF MASS COMMUNICATIONS

Beginning as far back as the 1920s, a number of smaller American companies, such as Shure Brothers and Electro-Voice, began to make significant contributions to microphone engineering and design. General applications, such as paging and sound reinforcement, required ingenious and economical solutions to many problems. Commercial development of capacitor microphones was more or less ruled out, due to the requirements for a cumbersome polarizing supply, so these companies concentrated primarily on moving coil designs.

The work of Bauer (1941) was significant in producing the Shure Unidyne directional (cardioid pattern) design based on a single moving coil element. Wiggins (1954) developed the Electro-Voice "Variable-D" single moving coil element, which provided low handling noise with excellent directional response.

Other companies designed crystal microphones for low cost, moderate-quality paging applications. These designs were based on the principle of *piezoelectricity* (from the Greek *piezien*, meaning pressure), which describes the property of many crystalline structures to develop a voltage across opposite facets when the material is bent or otherwise deformed. The early advantage of the piezos was a relatively high output signal, but eventually the coming of small, high energy magnet materials ruled them out.

THE GREAT CAPACITOR BREAKTHROUGH: THE ELECTRET

For many years the penalty carried by the capacitor microphone was its requirement for external polarization voltage. In the early sixties, Sessler and West of Bell Telephone Laboratories described a capacitor microphone which used a permanently polarized dielectric material between

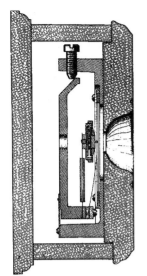

FIGURE 1-6

Blake's variable contact microphone.

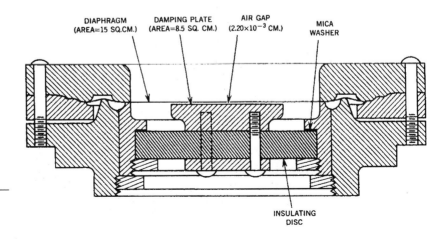

DIAPHRAGM (AREA=15 SQ.CM.)   DAMPING PLATE (AREA=8.5 SQ. CM.)   AIR GAP (2.20×10^{-3} CM.)   MICA WASHER

INSULATING DISC

FIGURE 1-7

The Wente capacitor microphone of 1917.

the movable diaphragm and the backplate of the microphone. Early materials exhibited significant losses in sensitivity over time, but these problems have been overcome. Further improvements have come in miniaturization, enabling electret microphones to be designed for a wide variety of close-in applications, such as tie-tack use, hidden on-stage pickup, and many other uses. Today's small electret microphone requires only a miniature self-contained 5-to-9 volt battery to power its equally miniature preamplifier. It is a testimony to the electret and its long-term stability and excellent technical performance that the Brüel & Kjaer series of studio microphones designed in the 1980s used electret technology.

STUDIO MICROPHONE TECHNOLOGY

The microphone is the first stage in the complex and extended technical chain between live performance and sound reproduction in the home or motion picture theater. Little wonder then that so much attention has been paid to the quality and technical performance of these fine instruments.

Capacitor microphones have dominated studio recording since the late 1940s when the first German and Austrian capacitor microphones came on the scene. As with any mature technology, progress comes slowly, and the best models available today have a useful dynamic range that exceeds that of a 20-bit digital recorder. With regard to directional performance, many newer microphones exhibit off-axis response integrity that far exceeds the best of earlier models.

At the beginning of the 21st century, it is interesting to observe the great nostalgia that many recording engineers have for the earlier vacuum tube capacitor models, especially the Neumann and AKG classic microphones of the 1950s. All of this reminds us that technology is so often tempered with subjective judgment to good effect.

THE FUTURE

The microphone per se is so highly developed that it is often difficult to see where specific improvements in the basic mechanisms are needed. Certainly in the area of increased directionality, via second and higher order designs, is there additional development engineering to be done. There may never be a direct-converting, high-quality digital microphone as such, but it is clear that digital signal processing will certainly play an important part in active microphone array development.

New usage concepts include microphones in conferencing systems, with their requirements for combining and gating of elements; and microphones in large arrays, where highly directional, steerable pickup patterns can be realized. These are among the many subjects that will be discussed in later chapters.

CHAPTER 2

BASIC SOUND TRANSMISSION AND OPERATIONAL FORCES ON MICROPHONES

INTRODUCTION

All modern microphones benefit from electrical amplification and thus are designed primarily to sample a sound field rather than take power from it. In order to understand how microphones work from the physical and engineering points of view, we must understand the basics of sound transmission in air. We base our discussion on sinusoidal wave generation, since sine waves can be considered the building blocks of most audible sound phenomena. Sound transmission in both plane and spherical waves will be discussed, both in free and enclosed spaces. Power relationships and the concept of the decibel are developed. Finally, the effects of microphone dimensions on the behavior of sound pickup are discussed.

BASIC WAVE GENERATION AND TRANSMISSION

Figure 2–1 illustrates the generation of a sine wave. The vertical component of a rotating vector is plotted along the time axis, as shown at A. At each 360° of rotation, the wave structure, or waveform, begins anew. The *amplitude* of the sine wave reaches a crest, or maximum value, above the zero reference baseline, and the *period* is the time required for the execution of one cycle. The term *frequency* represents the number of cycles executed in a given period of time. Normally we speak of frequency in *hertz* (Hz), representing cycles per second.

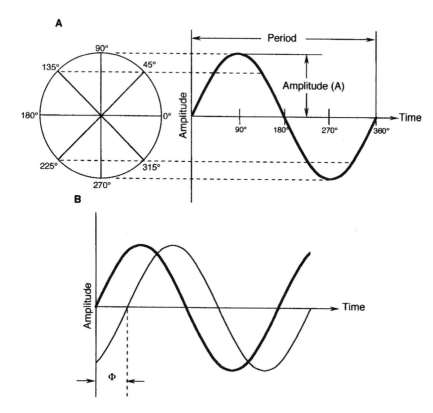

FIGURE 2-1 ——————

Generation of a sine
wave signal (A); phase
relationships between two
sine waves (B).

For sine waves radiating outward in a physical medium such as air, the baseline in Figure 2–1 represents the static atmospheric pressure, and the sound waves are represented by the alternating plus and minus values of pressure about the static pressure. The period then corresponds to *wavelength*, the distance between successive iterations of the basic waveform.

The speed of sound transmission in air is approximately equal to 344 meters per second (m/s), and the relations among speed (m/s), wavelength (m), and frequency (1/s) are:

$$c \text{ (speed)} = f \text{ (frequency)} \times \lambda \text{ (wavelength)}$$
$$f = c/\lambda$$
$$\lambda = c/f \qquad (2.1)$$

For example, at a frequency of 1000 Hz, the wavelength of sound in air will be 344/1000 = 0.344 m (about 13 inches).

Another fundamental relationship between two waveforms of the same frequency is their relative *phase* (ϕ), the shift of one period relative to another along the time axis as shown in Figure 2–1B. Phase is normally measured in degrees of rotation (or in *radians* in certain mathematical operations). If two sound waves of the same amplitude and

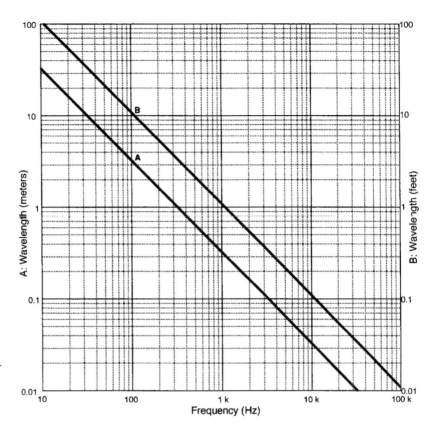

FIGURE 2–2

Wavelength of sound in air versus frequency; in meters (A); in feet (B).

frequency are shifted by 180° they will cancel, since they will be in an inverse (or anti-phase) relationship relative to the zero baseline at all times. If they are of different amplitudes, then their combination will not directly cancel.

The data shown in Figure 2–2 gives the value of wavelength in air when frequency is known. (By way of terminology, *velocity* and *speed* are often used interchangeably. In this book, speed will refer to the rate of sound propagation over distance, while velocity will refer to the specifics of localized air particle and air volume movement.)

TEMPERATURE DEPENDENCE OF SPEED OF SOUND TRANSMISSION

For most recording activities indoors, we can assume that normal temperatures prevail and that the effective speed of sound propagation will be as given above. There is a relatively small dependence of sound propagation on temperature, as given by the following equation:

$$\text{Speed} = 331.4 + 0.607\,°C \text{ m/s} \tag{2.2}$$

where °C is the temperature in degrees celsius.

ACOUSTICAL POWER

In any kind of physical system in which work is done, there are two quantities, one *intensive*, the other *extensive*, whose product determines the power, or rate at which work is done in the system. One may intuitively think of the intensive variable as the driving agent and the extensive variable as the driven agent. Table 2–1 may make this clearer.

Power is stated in *watts* (W), or joules/second. The *joule* is the unit of work or energy, and joules per second is the rate at which work is done, or energy expended. This similarity among systems makes it easy to transform power from one physical domain to another, as we will see in later chapters.

Intensity (I) is defined as power per unit area (W/m²), or the rate of energy flow per unit area. Figure 2–3 shows a sound source at the left radiating an acoustical signal of intensity I_0 uniformly into free space. We will examine only a small solid angle of radiation. At a distance of 10 m that small solid angle is radiating through a square with an area of 1 m², and only a small portion of I_0 will pass through that area. At a distance of 20 m the area of the square that accommodates the original solid angle is now 4 m², and it is now clear that the intensity at a distance of 20 m will be *one-fourth* what it was at 10 m. This of course is a necessary consequence of the law of conservation of energy.

TABLE 2–1   Intensive and extensive variables

System	Intensive variable	Extensive variable	Product
Electrical	voltage (e)	current (i)	watts (e × i)
Mechanical (rectilinear)	force (f)	velocity (u)	watts (f × u)
Mechanical (rotational)	torque (T)	angular velocity (θ)	watts (T × θ)
Acoustical	pressure (p)	volume velocity (U)	watts (p × U)

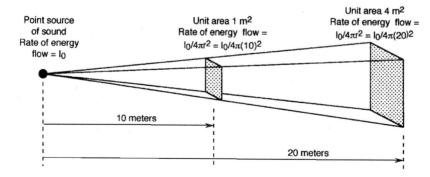

Point source of sound
Rate of energy flow = I_0

Unit area 1 m²
Rate of energy flow =
$I_0/4\pi r^2 = I_0/4\pi(10)^2$

Unit area 4 m²
Rate of energy flow =
$I_0/4\pi r^2 = I_0/4\pi(20)^2$

10 meters

20 meters

FIGURE 2–3

Sound intensity variation with distance over a fixed solid angle.

The relationship of intensity and distance in a free sound field is know as the inverse square law: as intensity is measured between distances of r and $2r$, the intensity changes from $1/I_0$ to $1/4I_0$.

The intensity at any distance r from the source is given by:

$$I = W/4\pi r^2 \tag{2.3}$$

The effective sound pressure in pascals at that distance will be:

$$p = \sqrt{I\rho_0 c} \tag{2.4}$$

where $\rho_0 c$ is the specific acoustical impedance of air (405 SI rayls).

For example, consider a point source of sound radiating a power of one watt uniformly. At a distance of 1 meter the intensity will be:

$$I = 1/4\pi(1)^2 = 1/4\pi = 0.08 \text{ W/m}^2$$

The effective sound pressure at that distance will be:

$$p = \sqrt{(0.08)405} = 5.69 \text{ Pa}$$

RELATIONSHIP BETWEEN AIR PARTICLE VELOCITY AND AMPLITUDE

The relation between air particle velocity (u) and particle displacement (x) is given by:

$$u(t) = j\omega \times (t) \tag{2.5}$$

where $\omega = 2\pi f$ and $x(t)$ is the maximum particle displacement value. The complex operator j produces a positive phase shift of $90°$.

Some microphones, notably those operating on the capacitive or piezoelectric principle, will produce constant output when placed in a constant amplitude sound field. In this case $u(t)$ will vary proportional to frequency.

Other microphones, notably those operating on the magnetic induction principle, will produce a constant output when placed in a constant velocity sound field. In this case, $x(t)$ will vary inversely proportional to frequency.

THE DECIBEL

We do not normally measure acoustical intensity; rather, we measure sound pressure level. One cycle of a varying sinusoidal pressure might look like that shown in Figure 2–4A. The peak value of this signal is shown as unity; the *root-mean-square* value (rms) is shown as 0.707, and the average value of the waveform is shown as 0.637. A square wave of unity value, shown at B, has *peak*, *rms*, and *average* values that all are unity. The *rms*, or effective, value of a pressure waveform corresponds directly to the power that is delivered or expended in a given acoustical system.

The unit of pressure is the pascal (Pa) and is equal to one newton/m². (The newton (N) is a unit of force that one very rarely comes across in

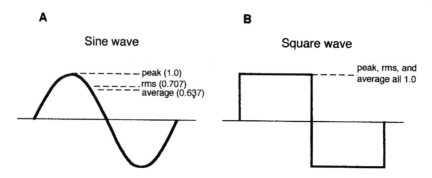

FIGURE 2–4

Sine (A) and square (B) waves: definitions of peak, rms and average values.

daily life and is equal to about 9.8 pounds of force.) Pressures encountered in acoustics normally vary from a low value of 20 μPa (micropascals) up to normal maximum values in the range of 100 Pa. There is a great inconvenience in dealing directly with such a large range of numbers, and years ago the decibel (dB) scale was devised to simplify things. The dB was originally intended to provide a convenient scale for looking at a wide range of power values. As such, it is defined as:

$$\text{Level (dB)} = 10 \log (W/W_0) \qquad (2.6)$$

where W_0 represents a reference power, say, 1 watt, and the logarithm is taken to the base 10. (The term *level* is universally applied to values expressed in decibels.) With one watt as a reference, we can say that 20 watts represents a level of 13 dB:

$$\text{Level (dB)} = 10 \log (20/1) = 13 \, \text{dB}$$

Likewise, the level in dB of a 1 milliwatt signal, relative to one watt, is:

$$\text{Level (dB)} = 10 \log (0.001/1) = -30 \, \text{dB}$$

From basic electrical relationships, we know that power is proportional to the *square* of voltage. As an analog to this, we can infer that acoustical power is proportional to the square of acoustical pressure. We can therefore rewrite the definition of the decibel in acoustics as:

$$\text{Level (dB)} = 10 \log (p/p_0)^2 = 20 \log (p/p_0) \qquad (2.7)$$

In sound pressure level calculations, the reference value, or p_0, is established as 0.00002 Pa, or 20 micropascals (20 μPa).

Consider a sound pressure of one Pa. Its level in dB is:

$$\text{dB} = 20 \log (1/0.00002) = 94 \, \text{dB}$$

This is an important relationship. Throughout this book, the value of 94 dB L_P will appear time and again as a standard reference level in microphone design and specification. (L_P is the standard terminology for sound pressure level.)

Figure 2–5 presents a comparison of a number of acoustical sources and the respective levels at reference distances.

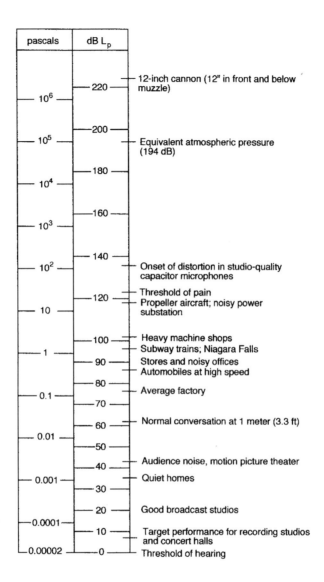

FIGURE 2-5

Sound pressure levels of various sound sources.

The graph in Figure 2–6 shows the relationship between pressure in Pa and L_P. The nomograph shown in Figure 2–7 shows the loss in dB between any two reference distances from a point source in the free field.

Referring once again to equation (2.4), we will now calculate the sound pressure level of one acoustical watt measured at a distance of 1 m from a spherically radiating source:

$$L_P = 20 \log (5.69/0.00002) = 109\,dB$$

It can be appreciated that one acoustical watt produces a considerable sound pressure level. From the nomograph of Figure 2–7, we can see that one acoustical watt, radiated uniformly and measured at a

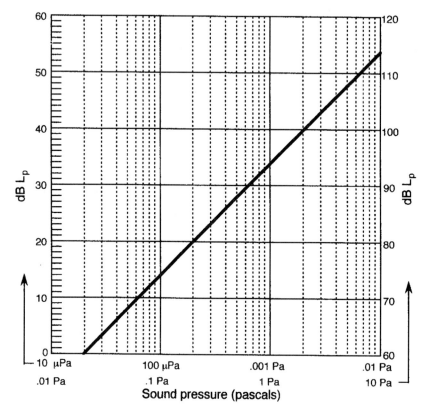

FIGURE 2–6

Relationship between sound pressure and sound pressure level.

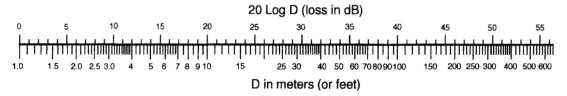

FIGURE 2–7

Inverse square sound pressure level relationships as a function of distance from the source; to determine the level difference between sound pressures at two distances, located the two distances and then read the dB difference between them; for example, determine the level difference between distances 50 m and 125 m from a sound source; above 50 read a level of 34 dB; above 125 read a level of 42 dB; taking the difference gives 8 dB.

distance of 10 m (33 feet), will produce $L_p = 89$ dB. How "loud" is a signal of 89 dB L_p? It is approximately the level of someone shouting in your face!

THE REVERBERANT FIELD

A free field exists only under specific test conditions. Outdoor conditions may approximate it. Indoors, we normally observe the interaction of a direct field and a reverberant field as we move away from a sound source. This is shown pictorially in Figure 2–8A. The reverberant field consists of the ensemble of reflections in the enclosed space, and reverberation time is considered to be that time required for the reverberant field to diminish 60 dB after the direct sound source has stopped.

There are a number of ways of defining this, but the simplest is given by the following equation:

$$\text{Reverberation time (s)} = \frac{0.16\,V}{S\overline{\alpha}} \tag{2.8}$$

where V is the room volume in m³, S is the interior surface area in m², and $\overline{\alpha}$ is the average absorption coefficient of the boundary surfaces.

The distance from a sound source to a point in the space where both direct and reverberant fields are equal is called *critical distance* (D_C). In live spaces critical distance is given by the following equation:

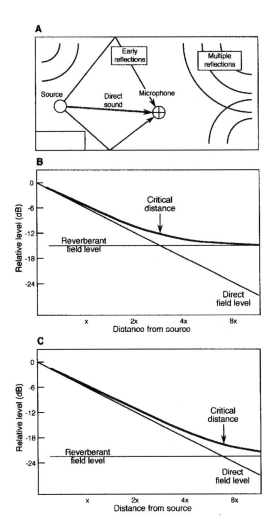

FIGURE 2-8

The reverberant field. Illustration of reflections in an enclosed space compared to direct sound at a variable distance from the sound source (A); interaction of direct and reverberant fields in a live space (B); interaction of direct and reverberant fields in a damped space (C).

$$D_C = 0.14\sqrt{QS\bar{\alpha}} \qquad\qquad (2.9)$$

where Q is the *directivity factor* of the source. We will discuss this topic in further detail in Chapter 17.

In a live acoustical space, $\bar{\alpha}$ may be in the range of 0.2, indicating that, on average, only 20% of the incident sound power striking the boundaries of the room will be absorbed; the remaining 80% will reflect from those surfaces, strike other surfaces, and be reflected again. The process will continue until the sound is effectively damped out. Figures 2–8B and C show, respectively, the observed effect on sound pressure level caused by the interaction of direct, reflected, and reverberant fields in live and damped spaces.

Normally, microphones are used in the direct field or in the transition region between direct and reverberant fields. In some classical recording operations, a pair of microphones may be located well within the reverberant field and subtly added to the main microphone array for increased ambience.

SOUND IN A PLANE WAVE FIELD

For wave motion in a free plane wave field, time varying values of sound pressure will be in phase with the air particle velocity, as shown in Figure 2–9. This satisfies the conditions described in Table 2.1, in which the product of pressure and air volume velocity define acoustical power. (Volume velocity may be defined here as the product of particle velocity and the area over which that particle velocity is observed.)

If a microphone is designed to respond to sound pressure, the conditions shown in Figure 2–9A are sufficient to ensure accurate reading of the acoustical sound field.

Most directional microphones are designed to be sensitive to the air pressure difference, or *gradient*, existing between two points along a given pickup axis separated by some distance l. It is in fact this sensitivity that enables these microphones to produce their directional pickup characteristics. Figure 2–9B shows the phase relationships at work here. The pressure gradient [dp/dl] is in phase with the particle displacement [x(t)]. However, the particle displacement and particle velocity [dx/dt] are at a 90° phase relationship.

These concepts will become clearer in later chapters in which we discuss the specific pickup patterns of directional microphones.

SOUND IN A SPHERICAL WAVE FIELD

Relatively close to a radiating sound source, the waves will be more or less spherical. This is especially true at low frequencies, where the difference in wavefront curvature for successive wave crests will be quite pronounced. As our observation point approaches the source, the phase

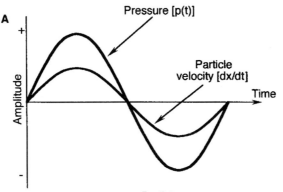

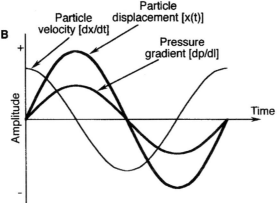

FIGURE 2-9

Wave considerations in microphone performance: relationship between sound pressure and particle velocity (A); relationship among air particle velocity, air particle displacement, and pressure gradient (B); relationship between pressure and pressure gradient (C) (Data presentation after Robertson, 1963).

angle between pressure and particle velocity will gradually shift from zero (in the far field) to 90°, as shown in Figure 2–10A. This will cause an increase in particle velocity with increasing phase shift, as shown at *B*.

As we will see in a later detailed discussion of pressure gradient microphones, this phenomenon is responsible for what is called *proximity effect*, the tendency of directional microphones to increase their LF (low frequency) output at close operating distances.

EFFECTS OF HUMIDITY ON SOUND TRANSMISSION

Figure 2–11 shows the effects of both inverse square losses and HF losses due to air absorption. Values of relative humidity (RH) of 20% and 80% are shown here. Typical losses for 50% RH would be roughly halfway between the plotted values shown.

For most studio recording operations HF losses may be ignored. However, if an organ recording were to be made at a distance of 12 m in a large space and under very dry atmospheric conditions, the HF losses could be significant, requiring an additional HF boost during the recording process.

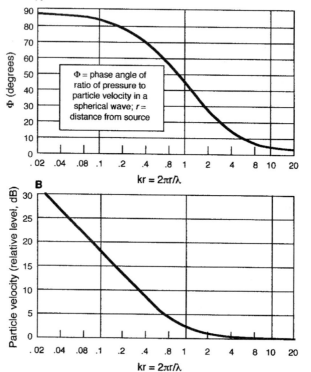

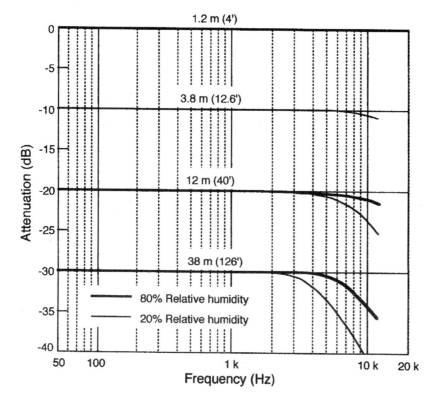

FIGURE 2-10

Spherical sound waves: phase angle between pressure and particle velocity in a spherical wave at low frequencies; r is the observation distance and λ is the wavelength of the signal (A); increase in pressure gradient in a spherical wave at low frequencies (B).

FIGURE 2-11

Effects of both inverse square relationships and HF air losses (20% and 80% RH).

DIFFRACTION EFFECTS AT SHORT WAVELENGTHS; DIRECTIVITY INDEX (DI)

Microphones are normally fairly small so that they will have minimal effect on the sound field they are sampling. There is a limit, however, and it is difficult to manufacture studio quality microphones smaller than about 12 mm (0.5 in) in diameter. As microphones operate at higher frequencies, there are bound to be certain aberrations in directional response as the dimensions of the microphone case become a significant portion of the sound wavelength. *Diffraction* refers to the bending of sound waves as they encounter objects whose dimensions are a significant portion of a wavelength.

Many measurements of off-axis microphone response have been made over the years, and even more theoretical graphs have been developed. We will now present some of these.

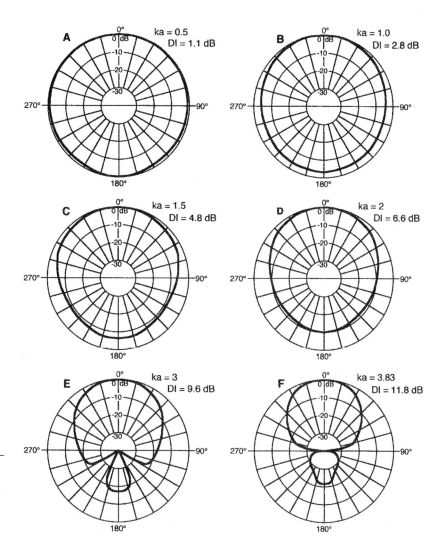

FIGURE 2-12

Theoretical polar response for a microphone mounted at the end of a tube. (Data presentation after Beranek, 1954.)

Figure 2–12 shows polar response diagrams for a circular diaphragm at the end of a long tube, a condition that describes many microphones. In the diagrams, $ka = 2\pi a/\lambda$, where a is the radius of the diaphragm. Thus, ka represents the diaphragm circumference divided by wavelength. *DI* stands for *directivity index*; it is a value, expressed in decibels, indicating the ratio of on-axis pickup relative to the total pickup integrated over all directions. Figure 2–13 shows the same set of measurements for a microphone which is effectively open to the air equally on both sides. It represents the action a ribbon microphone, with its characteristic "figure-eight" angular response.

Figure 2–14 shows families of on- and off-axis frequency response curves for microphones mounted on the indicated surfaces of a cylinder and a sphere. Normally, a limit for the HF response of a microphone would be a diameter/λ ratio of about one.

In addition to diffraction effects, there are related response aberrations due to the angle at which sound impinges on the microphone's diaphragm. Figure 2–15A shows a plane wave impinging at an off-axis oblique angle on a microphone diaphragm subtended diameter which is one-fourth of the sound wavelength. It can be seen that the center portion of the diaphragm is sampling the full value of the waveform, while adjacent portions are sampling a slightly lesser value. Essentially, the diaphragm will respond accurately, but with some small diminution of output for the off-axis pickup angle shown here.

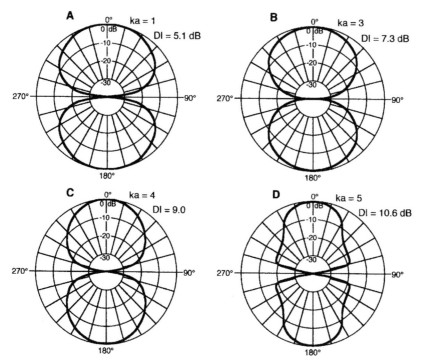

FIGURE 2–13

Theoretical polar response for a free microphone diaphragm open on both sides. (Data presentation after Beranek, 1954.)

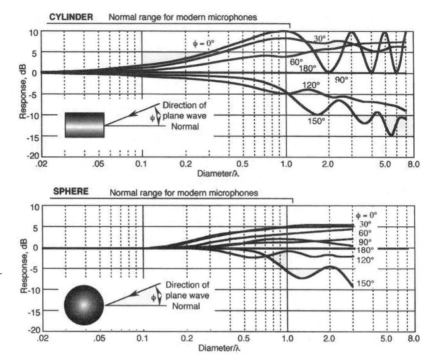

FIGURE 2–14

On and off-axis frequency response for microphones mounted on the end of a cylinder and a sphere. (Data after Muller et al., 1938.)

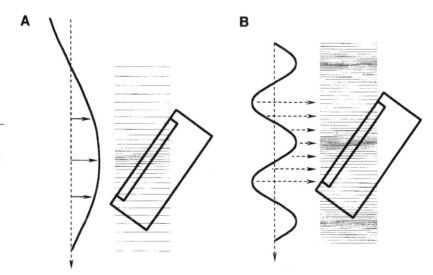

FIGURE 2–15

Plane sound waves impinging on a microphone diaphragm at an oblique angle. Microphone diaphragm subtended diameter equal to $\lambda/4$ (A); microphone diaphragm subtended diameter equal to λ (B). (Data after Robertson, 1963.)

The condition shown in Figure 2–15B is for an off-axis sound wavelength which is equal to the subtended diameter of the microphone diaphragm. Here, the diaphragm samples the entire wavelength, which will result in near cancellation in response over the face of the diaphragm.

C H A P T E R   3

THE PRESSURE MICROPHONE

INTRODUCTION

An ideal pressure microphone responds only to sound pressure, with no regard for the directional bearing of the sound source; as such, the microphone receives sound through a single active opening. In reality, a pressure microphone exhibits some directionality along its main axis at short wavelengths, due principally to diffraction effects. As we saw in Figure 2–12, only as the received wavelength approaches the circumference of the microphone diaphragm does the microphone begin to depart significantly from omnidirectional response. For many studio-quality pressure microphones, this becomes significant above about 8 kHz.

The earliest microphones were of the pressure type, and the capacitor pressure microphone is widely used today in music recording as well as in instrumentation and measurement applications. By way of terminology, the *capacitor* microphone is familiarly referred to in the sound industry as the *condenser* microphone, making use of the earlier electrical term for capacitor. We use the modern designation capacitor throughout this book.

We begin with a study of the capacitor pressure microphone, analyzing it in physical and electrical detail. We then move on to a similar analysis of the dynamic pressure microphone. Other transducers that have been used in pressure microphone design, such as the piezoelectric effect and the loose contact (carbon granule) effect, are also discussed. The RF (radio frequency) signal conversion principle use in some capacitor microphones is discussed in Chapter 8.

ANALYSIS OF THE CAPACITOR PRESSURE MICROPHONE

Figure 3–1 shows section and front views of a modern studio capacitor pressure microphone capsule along with its associated electrical circuitry. It is very similar to Wente's original 1917 model (see Figure 1–7), except that the modern design shown here is about one-third the diameter of the

Wente model. General dimensions and relative signal values for a typical 12.7 mm (0.5 in) diameter capacitor pressure microphone are:

1. Dimension h is about 20 μm (0.0008 in).
2. Peak-to-peak diaphragm displacement for a 1 Pa rms sine wave signal (94 dB L_P) is about 1×10^{-8} m.
3. The static capacitance of the capsule is about 35 pF.
4. For a dc polarizing voltage on the diaphragm of 60 V, the signal voltage generated by the 1 Pa acoustical signal is about 12 Vrms at the capacitor's terminals before preamplification.

Table 3–1 indicates the peak diaphragm displacement as a function of pressure level. For the sake of example, let us multiply the microphone's dimensions by 1 million. The microphone now is 12 km in diameter, and

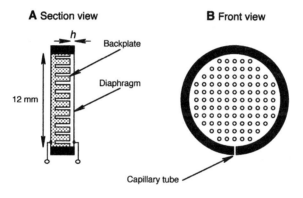

A Section view **B** Front view

Backplate

Diaphragm

12 mm

Capillary tube

C Electrical circuit

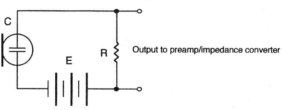

C

E

R

Output to preamp/impedance converter

FIGURE 3-1

Details of a 12 mm diameter capacitor microphone: section view (A); front view (B); simplified electrical circuit (C).

TABLE 3-1   Peak diaphragm displacement relative to pressure level

Level (dB L_P)	Peak-to-peak displacement (m)
14	10^{-12}
34	10^{-11}
54	10^{-10}
74	10^{-9}
94	10^{-8}
114	10^{-7}
134	10^{-6}

the distance from the backplate to the diaphragm is 20 m. For L_P of 94 dB, the peak-to-peak displacement of the diaphragm is about 10 mm. Using the same model, the peak diaphragm displacement at 134 dB L_P is about 1 m, representing a ±5% variation in spacing between the back-plate and diaphragm. This exercise aptly demonstrates the microscopic nature of the capacitor microphone's diaphragm motion at normal sound levels.

PHYSICAL DETAILS OF THE CAPACITOR PRESSURE MICROPHONE

The very small holes in the backplate (Figure 3–1B) are evenly distrib-uted on a uniform grid. During the actual back and forth motion of the diaphragm, the air captured in the holes provides damping of the diaphragm's motion at its principal resonance, which is normally in the range of 8–12 kHz. The diaphragm is usually made of a thin plastic material, typically Mylar, on which has been deposited a molecular-thin layer of metal, often gold. Aluminum, nickel, and titanium also have been used as diaphragm materials.

Operating in parallel with the tension of the diaphragm itself is the added stiffness provided by the captured air behind the diaphragm. Both of these are necessary to maintain the high resonance frequency of the diaphragm assembly. A very small capillary tube connects the interior air mass to the outside, providing a slow leakage path so that static atmos-pheric pressure will equalize itself on both sides of the diaphragm under all atmospheric conditions. The polarizing circuitry is shown at C.

One early notable capacitor microphone model, the Altec-Lansing 21C, used a very thin 50 μm (0.002 in) gold sputtered glass plate in place of the normal flexible diaphragm. The inherent stiffness of the glass plate required no additional tension to attain a suitably high-resonance frequency.

DIAPHRAGM MOTION

While we customarily speak of a capacitor microphone's diaphragm as having a single degree of freedom, the actual motion over most of the frequency range resembles that of an ideal drum head, as shown in exaggerated form in Figure 3–2A. It is intuitively clearer and mathe-matically simpler to think of the diaphragm as a rigid piston. If we take the center of the diaphragm as our reference displacement value, it is obvious that the outer portions of the diaphragm exhibit less displace-ment. Wong and Embleton (1994) show that the effective area of an equivalent rigid piston is one that has an area *one-third* that of the actual diaphragm.

As the circular edge-supported membrane vibrates at higher fre-quencies, its motion no longer is simple. Motions influenced by radial and tangential modes in the diaphragm, such as those shown in Figures 3–2B and 3–2C begin to appear, and the microphone's amplitude

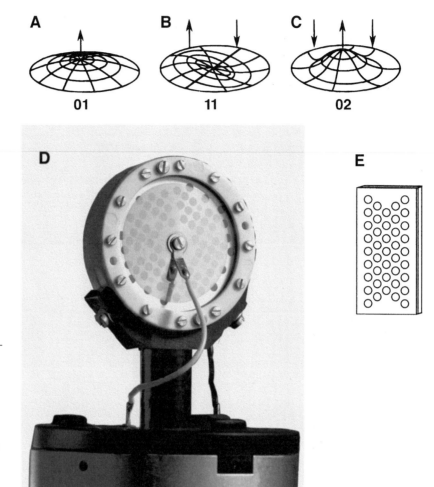

FIGURE 3–2

Complex diaphragm
vibration due to radial and
tangential modes: normal
motion (A); mode 11 (B);
mode 02 (C); Neumann
center-clamped assembly
(D); rectangular assembly as
used by Pearl Microphone
Laboratories (E). (Photo
courtesy Neumann/USA.)

response becomes erratic. As a general rule, radial modes become pre-
dominant only above the normal frequency range of the microphone.

Figure 3–2D shows details of Neumann's back-to-back diaphragm
assembly with its characteristic center and rim clamping arrangement.
Obviously, this diaphragm has a completely different set of motions than
those of the simple rim supported diaphragm. Figure 3–2E shows a rectangular diaphragm structure, with yet another set of motions at high
frequencies. A diaphragm of this type is normally about 12 mm (0.5 in)
wide and about 38 mm (1.5 in) high.

An important design variation is used by Sennheiser Electronics, and
that is the practice of *not* increasing the diaphragm tension to maintain
flat response at high frequencies. Instead, flat response is obtained by
electrical boosting of the diaphragm's output at HF to whatever degree
is required.

ELECTRICAL DETAILS OF THE CAPACITOR PRESSURE MICROPHONE

The capacitor microphone operates on the static relationship

$$Q = CE \qquad (3.1)$$

where Q is the electrical charge (coulombs) on the plates of a capacitor, C is the capacitance (farads), and E is the applied dc voltage. In the capacitor microphone, the backplate and diaphragm represent the two plates of the capacitor.

Figure 3–3 shows a capacitor with variable spacing between the two plates. The capacitor at A is charged by applying a reference dc voltage, E, across it; and a charge, Q, is produced on the plates. Now, if the plates are separated slightly, as shown at B, the capacitance is reduced. The charge remains fixed, however, and this causes the voltage, E, to rise. This comes as a consequence of satisfying equation (3.1). Alternatively, if we move the plates closer together, as shown at C, the capacitance increases and the voltage drops.

In the externally charged capacitor microphone, the applied polarizing voltage is fed through a very high resistance, as in the network shown in Figure 3–1C. The series resistance of 1 GΩ (10^9 ohms) ensures that, once the charge is in place, normal audio frequency variations in the diaphragm's displacement do not alter the charge but are manifest as a variation in voltage, which is the inverse of the change in capacitance. For our analysis here, we assume that negligible power is transferred between the capacitor and the resistor.

It is important that the value of the polarizing voltage be well below the break-down voltage of the air dielectric in the capacitor capsule. This value is approximately equal to 3000 V/mm under normal atmospheric conditions. Studio-grade capacitor microphones typically are polarized in the range of 48–65 volts, while instrumentation microphones may operate in the range of 200 V. A slight electrostatic attraction between the diaphragm and backplate is caused by the polarizing voltage, which causes the diaphragm to be drawn very slightly toward the backplate.

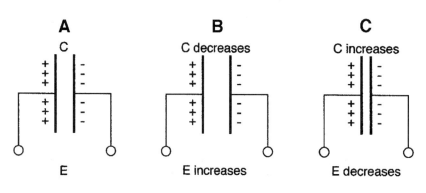

FIGURE 3–3

Relationships in a charged capacitor: a fixed charge is placed on the plates of a capacitor (A); reducing the capacitance causes the voltage to increase (B); increasing the capacitance causes the voltage to decrease (C).

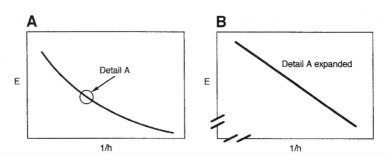

FIGURE 3–4

Nonlinearity of capacitance versus displacement (A); for a very small displacement, the relationship is virtually linear (B).

For a fixed charge, the basic polarizing voltage versus capacitance relationship is not truly a linear one. When viewed over a wide range of capacitance change, the relation is hyperbolic, as shown in Figure 3–4A. However, over the normally small displacement range, as shown at *B*, the relationship is very nearly linear. Recall from the section above on Analysis of the Capacitor Pressure Microphone that normal diaphragm motion is extremely small; only for signals in the range of 130 dB L_P and greater is the diaphragm displacement in the 5% range relative to its distance from the backplate, and this represents a change in position of only 1 part in 20.

Shielding the microphone is very important since the high electrical impedance of the polarizing network makes the system susceptible to electrostatic interference. Normally, a grounded metal mesh screen separates the diaphragm assembly from the outside and provides the necessary shielding.

REGION OF FLAT ELECTRICAL OUTPUT

The value of capacitance depends on, among other things, the displacement of one plate relative to the other. Therefore, for uniform (flat) electrical output from the capacitor element, the diaphragm displacement should be independent of frequency. In a plane progressive wave, pressure and particle velocity are in phase with each other, as discussed in Chapter 2. The relationship between air particle displacement and velocity is given by the integral of $u(t)$ with respect to time:

$$x(t) = \int u(t)dt \qquad (3.2)$$

where $x(t)$ is the instantaneous particle displacement and $u(t)$ is the instantaneous particle velocity. For a sinusoidal signal, $u(t)$ is of the form $e^{j\omega t}$; therefore,

$$x(t) = \int u(t)dt = (-j/\omega)e^{j\omega t} \qquad (3.3)$$

This equation describes a response that is inversely proportional to frequency (i.e., rolls off at 6 dB per octave with increasing frequency).

The response is further modified by the $-j$ complex operator, which shifts the relative phase by $-90°$.

Remember that the diaphragm is under considerable mechanical tension, with additional stiffness provided by the small air chamber behind it. When a pressure wave impinges on the diaphragm, it encounters a mechanical responsiveness proportional to $j\omega/S$, where S is the mechanical stiffness of the diaphragm (N/m). The $j\omega/S$ term describes a response directly proportional to frequency (i.e., rises 6 dB per octave with increasing frequency). The response is further modified by the positive j complex operator, which shifts the relative phase by 90°. We see the two effects are complementary and their combined effect is shown as:

$$(-j/\omega)e^{j\omega t} \times (j\omega/S) = e^{j\omega t}/S \qquad (3.4)$$

where $e^{j\omega t}/S$ now is of the same form as $u(t)$.

Therefore, the response is flat and the net phase shift is 0° in the range over which the diaphragm is stiffness controlled.

The entire process is shown graphically in Figure 3–5. At A, the free field relationships among air particle velocity, pressure, and displacement are shown. At B, the combination of air particle displacement and diaphragm resonance is shown.

Between the ranges of diaphragm stiffness and mass control is the region of resonance; and for most studio microphones, this ranges from about 8–12 kHz. In most microphones, the resonance is fairly well damped so as not to cause an appreciable rise in response; perhaps no more than about 2 or 3 dB. Beyond resonance, the overall response of the microphone begins to roll off at a rate of 12 dB per octave; however, the on-axis response of the microphone tends to remain fairly flat over about half an octave above resonance, due to diffraction effects similar to those shown in Figure 2–14.

The output of the capacitor pressure microphone remains very flat down to the lower frequencies, limited only by the movement of air through the atmosphere pressure adjusting effect of the capillary tube and the variations in charge on the capacitor plates through the very high

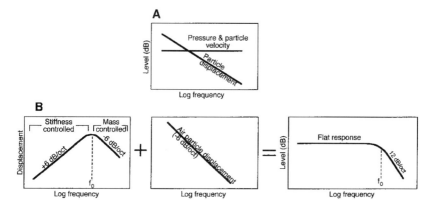

FIGURE 3–5

Forces on the diaphragm: relations among pressure, air particle velocity, and air particle displacement (A); response of a stiffness-controlled diaphragm (B).

biasing resistor (see Figure 3–1). In most studio-quality capacitor pressure microphones, LF rolloff effects are usually seen only in the range of 10 Hz or lower.

CAPACITOR MICROPHONE SENSITIVITY TO TEMPERATURE AND BAROMETRIC PRESSURE

For normal recording, broadcasting, and sound reinforcement applications using capacitor microphones, changes in response due to temperature and barometric pressure variations generally may be ignored. However, in many instrumentation applications, the variations may be significant enough to warrant recalibration of the measurement system.

The primary effect of temperature increase is a reduction of diaphragm tension, which causes an increase in sensitivity and a decrease in bandwidth. The net effect on sensitivity is quite small, roughly of the order of -0.005 dB/°C. (See equations (3.7) and (3.8).)

A decrease in barometric pressure has a slight effect on the LF and MF (medium frequency) sensitivity of the microphone, but the decrease in air pressure at the diaphragm at f_0 produces less damping of the diaphragm motion at resonance, causing an increase in the response at f_0.

The effects of both temperature and atmospheric pressure variations on the response of an instrumentation microphone are shown in Figures 3–6A and 3–6B, respectively.

EQUIVALENT ELECTRICAL CIRCUITS FOR THE CAPACITOR PRESSURE MICROPHONE

Although the capacitor pressure microphone is of relatively simple construction, it consists of many individual acoustical masses, compliances, and resistances. Figure 3–7 shows a complete circuit with definitions of all elements. The transformer represents the conversion from the acoustical domain into the electrical domain.

At LF and MF the circuit can be simplified as shown in Figure 3–8. At HF, the circuit can be simplified as shown in Figure 3–9A, and the response curves in Figure 3–9B show the effect of varying the damping on the diaphragm by altering the value of R_{AS}.

DETAILS OF THE PREAMPLIFIER AND POLARIZING CIRCUITRY

Figure 3–10A shows details of the microphone preamplifier and polarizing system for a vacuum tube design. Much of the complexity here is determined by the external powering source for polarization and the necessity for correct dc biasing of the vacuum tube.

Figure 3–10B shows a similar arrangement for a modern FET (field-effect transistor) solid-stage design. Note here the presence of a 10 dB attenuating (padding) capacitor around the capsule; the switchable capacitor shunts the signal generated by the capsule so that the system

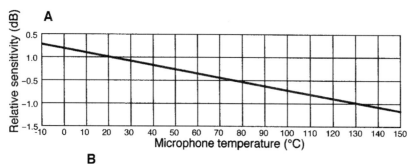

A

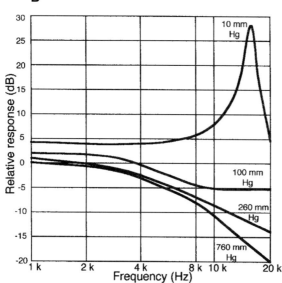

B

FIGURE 3-6

Effects of temperature
variations on the sensitivity
of a capacitor pressure
microphone (A); effect of
atmospheric pressure
variations on the response
of a capacitor pressure
microphone (B). (Data after
Brüel and Kjaer, 1977.)

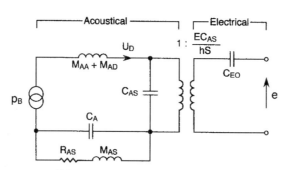

Where:

p_B = rms pressure at diaphragm
M_{AA} = air mass associated with diaphragm
M_{AD} = acoustic mass of diaphragm
U_D = rms volume velocity of diaphragm
C_{AS} = compliance of diaphragm
S = effective area of diaphragm
C_A = compliance of air behind the diaphragm
R_{AS} = resistance of air in holes in backplate
M_{AS} = mass of air in holes in backplate
C_{EO} = electrical capacitance measured with force $f = 0$
e = output voltage

FIGURE 3-7

Equivalent electroacoustical
circuit for a capacitor
pressure microphone
(impedance analogy).
(Data after Beranek,
1954.)

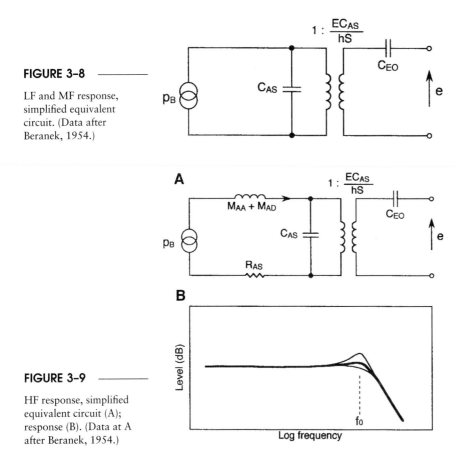

FIGURE 3–8

LF and MF response, simplified equivalent circuit. (Data after Beranek, 1954.)

FIGURE 3–9

HF response, simplified equivalent circuit (A); response (B). (Data at A after Beranek, 1954.)

can operate at higher sound levels without causing electrical overload in the following preamplification stage.

The shunt capacitor is normally chosen to have a value of about 0.4 that of the capsule itself. The parallel combination of the two capacitors then has a value of about one-third that of the capsule itself, resulting in an attenuation of output of about 10 dB. The action of the shunt capacitor is not like that of a simple voltage divider; rather, a slight signal non-linearity is caused by the circuit.

Let x represent a variable corresponding to the normal signal incident on the capsule. Then xC represents the varying value of capsule capacitance without the shunt. The parallel combination of the capsule and the shunt has a net capacitance equal to

$$C_{net} = \frac{x \times 0.4C^2}{0.4C + Cx}$$

which is of the form

$$C_{net} = \frac{Ax}{B + Cx} \qquad (3.5)$$

where $A = 0.4C^2$, $B = 0.4C$, and C is the capacitance of the capsule.

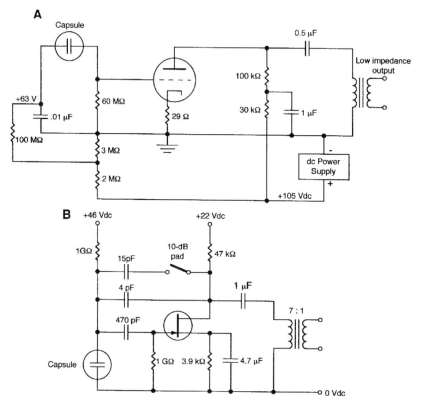

FIGURE 3-10 ————

Details of capacitor
microphone preamplifiers:
vacuum tube type (A);
solid-state type (B).

Equation 3.5 can be expanded to

$$C_{net} = \frac{ABx + ACx^2}{B^2 - (Cx)^2} \qquad (3.6)$$

In this form, we see that the simple relationship Cx, which represents the variable capacitance of the unattenuated capsule, has been transformed into a combination that includes squared terms in both numerator and denominator, indicating the presence of second harmonic distortion.

The effect of this is small, considering the minute signal variations represented by x. In any event, the distortion introduced by the shunt capacitance is negligible under the high sound pressure operating circumstances that would normally call for its use.

Signal output at the capsule can also be attenuated by dropping the dc bias voltage padding in this manner. The Neumann models TLM170 and TLM50 achieve their capsule padding in this manner. The TLM50 is padded by changing the value of dc bias voltage on the capsule from 60 to 23 V. This causes a slight alteration in the microphone's HF response due to the reduced electrostatic attraction between diaphragm and backplate, which causes the diaphragm to move slightly farther away from the backplate.

Regarding the output stage of the microphone's built-in preamplifier, the typical electrical output impedance is in the range of 50–200 ohms. The output is normally balanced, either with a transformer or by balanced solid-state output circuitry, to ensure interference-free operation over long transmission paths downstream. The electrical load normally seen by the microphone preamplifier is in the range of 1500–3000 ohms, which approximates essentially an unloaded condition if the driving source has a low impedance. Microphone wiring runs with low capacitance cable may extend up to 200 m with negligible effect on response.

ON-AXIS VERSUS RANDOM INCIDENCE RESPONSE: GRIDS, BAFFLES, AND NOSE CONES

As we saw in Figure 2–14, a free progressive plane wave arriving along the primary axis of a cylinder or a sphere shows a considerable rise in HF response. For the cylinder, the maximum on-axis rise will be 10 dB relative to the response at 90°; by comparison, the sphere produces only a 6 dB on-axis rise. In both cases, the response at 90° is very nearly flat and, in fact, approximates the integrated response of the microphone in a random sound field.

In the design of pressure microphones for both instrumentation and recording the diffuse random field. The smaller the diameter of the obstacle, the higher in frequency the divergence between random and on-axis response, as shown in Figure 3–11. Here, the microphones have been designed for essentially flat on-axis response.

FIGURE 3–11

Typical on-axis and random response of 12 mm (05 in) and 25 mm (1 in) pressure microphones.

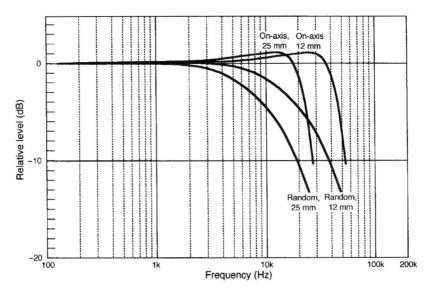

Figure 3–12 shows how a Brüel & Kjaer 4000 series omnidirectional microphone can be modified in HF response by changing the protective grid. The data shown at A is for the normal version of the microphone, with flat on-axis response (−6 dB at 40 kHz) and rolled-off random incidence response. When the normal grid is replaced with an alternate grid that produces less diaphragm damping, the response in panel B is produced. Here, the random incidence response is flat to 15 kHz and the on-axis response exhibits a HF peak of about 6 dB.

The same microphone, with its normal grid replaced by a nose cone baffle providing indirect access to the diaphragm, produces the response shown at C for all directions of sound incidence.

The Neumann M50 microphone, designed during the mid-1900s, consists of a 12 mm diameter capacitor element mounted on a sphere

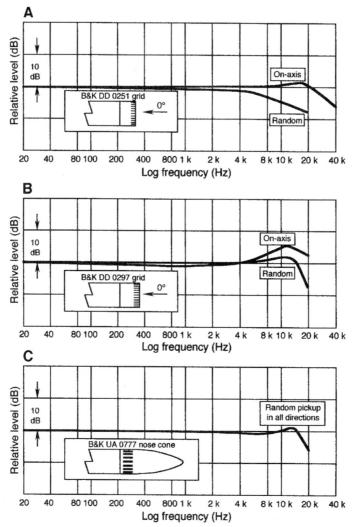

FIGURE 3–12

Microphones designed for specific pickup characteristic: flat on-axis response with rolled-off random incidence response (A); flat random incidence with peaked response on-axis (B); use of a special nose cone baffle to produce random incidence response in all pickup directions (C). (Data after Brüel and Kjaer, 1977.)

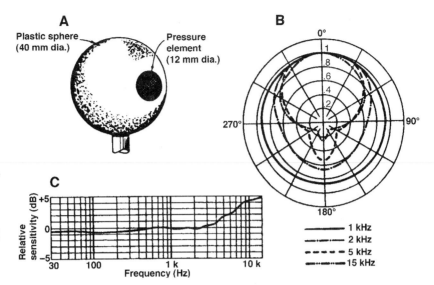

FIGURE 3-13

Neumann M50 data,
including sketch of
capacitor capsule mounting
(A); polar response (B); and
on-axis frequency response
(C). (Figure courtesy
Neumann/USA.)

approximately 40 mm in diameter, approximating the spherical response shown in Figure 2–14 for a diameter/wavelength corresponding to about 8200 Hz. Note the very close agreement between the on-axis response of the M50 (Figure 3–13) and the data shown in Figure 2–14, when the normalized unity value on the frequency axis is equated to 8200 Hz.

The M50 exhibits a flat diffuse field response but depends on the sphere to give added HF presence to signals arriving on-axis. This microphone has never lost its popularity with classical recording engineers and often is used in the transition zone between the direct and reverberant fields for added presence at high frequencies.

The Sennheiser MKH20, a flat on-axis pressure microphone, can be converted to essentially flat random incidence response by engaging an internal electrical HF shelving boost. It also can be given preferential on-axis directivity by the addition of a small rubber ring mounted at the end of the microphone. Details are shown in Figure 3–14. (The Sennheiser MKH20 employs a radio frequency signal conversion system, which we will discuss in Chapter 8. In terms of directional response, RF capacitor microphones have the same characteristics as the other capacitor models discussed in this chapter.)

Overall, the effects of diffraction and off-axis wave interference at the diaphragm of a capacitor pressure microphone placed at the end of a 21 mm diameter body produces the polar response versus frequency as shown in Figure 3–15.

TYPICAL CAPACITOR PRESSURE MICROPHONE NOISE SPECTRA

The capacitor pressure microphone represents a careful balance of technical attributes. It reached an advanced level of overall performance

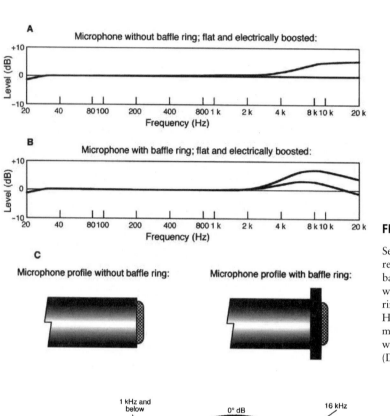

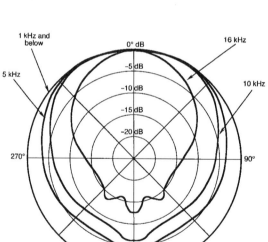

FIGURE 3–14 ——————

Sennheiser MKH20 data:
response with and without
baffle ring (A); response
with and without baffle
ring, electrically boosted at
HF (B); profiles of
microphone with and
without baffle ring (C).
(Data courtesy Sennheiser.)

FIGURE 3–15 ——————

Set of polar curves for
a pressure microphone
placed at the end of a
21 mm diameter cylinder,
exhibiting narrowing of
response patterns at HF.
(Data after Boré, 1989.)

during the middle 1980s, and today's many models invariably represent
different "operating points" among the variables of low noise, distortion
at high levels, bandwidth extension, and polar pattern control.

With the coming of digital recording, the self-noise of a capacitor
capsule and its associated preamplifier have come under close scrutiny,
since the effective noise floor of the Compact Disc and other higher

density formats is vastly lower than that of the long-playing stereo disc. Microphone noise levels that were acceptable in the predigital era may not be acceptable today. Over the years, we have seen the weighted noise floors of studio-quality capacitor microphones drop by 10–12 dB.

The self-noise floor of a capacitor microphone is normally weighted according to the *A curve* shown in Figure 3–16 and stated as an equivalent acoustical rating. For example, a noise rating of 10 dB(A) indicates that the noise floor of the microphone is approximately equivalent to that which a theoretically perfect microphone would pick up if it were operating in an actual acoustical space that had a residual noise rating of 10 dB(A), or its equivalent value of *NC 10*.

These matters are discussed in detail in Chapter 7, which is devoted to microphone measurements; here, we intend only to detail the spectral nature of the noise itself.

Figure 3–17 shows the composite one-third octave noise spectra of a preamplifier with input capacitances equivalent to 25 mm (1 in) and 12 mm (0.5 in) instrumentation microphones. Note that as the diaphragm diameter decreases by one half the noise floor rises approximately 6 dB. However, as a tradeoff of performance attributes, each

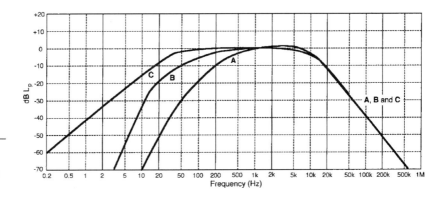

FIGURE 3–16

Standard weighting curves for acoustical noise measurements.

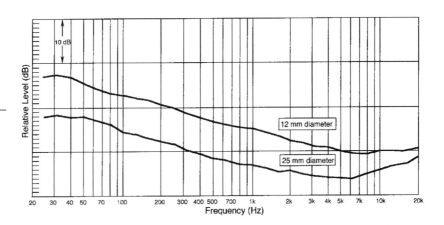

FIGURE 3–17

One-third octave noise spectra at the output of a microphone preamplifier with input loaded by capacitors equivalent to 25 mm and 12 mm diameter capsules.

succeeding halving of diameter produces a microphone with *twice* the HF bandwidth capability of the preceding one. Fortunately, the ear's sensitivity to low frequencies is diminished at low levels, as indicated by the weighting curves shown in Figure 3–16.

The curves show a rise below about 1 kHz of approximately 10 dB per decade with decreasing frequency. This rise is usually referred to as *1/f* noise (inversely proportional to frequency) and is primarily a property of the electronics. The flattening of the spectrum and eventual slight upturn at HF largely is the result of the capacitor element itself and the reflection of its mechano-acoustical resistance into the electrical domain.

Newer electronics have reduced *l/f* noise considerably; however, this is a minor improvement when we consider the relative insensitivity of the ear to low frequencies at low levels. Typical microphone preamplifier output noise spectra are shown in Figure 3–17.

THE ELECTRET CAPACITOR PRESSURE MICROPHONE

Electret materials have been known for more than a century, but only in the last 35 years or so has the electret had an impact on capacitor microphone design, bringing excellent performance to very low-cost models. The electret is a prepolarized material, normally polytetrafluoroethylene, which has been given a permanent electrostatic charge through placement in a strong electric field and under heat. As the heat is withdrawn, the electric charge remains. The material is virtually the electrostatic equivalent of a permanent magnet, and if a capacitor backplate is coated with one of the newer electret materials, the resulting microphone will have the same performance characteristics as a standard capacitor capsule with an effective polarizing value of about 100 V. Figure 3–18A shows a section view of a capacitor capsule using an electret backplate.

Alternatively, an electret diaphragm can be used, as shown in Figure 3–18B. One drawback here is that the metallized electret foil has somewhat greater mass per unit area than typical non-electret diaphragm materials, possibly compromising HF response.

Figure 3–19A shows a typical electrical implementation of the electret; note the simplicity of the design. A photograph of a small electret microphone is shown in Figure 3–19B.

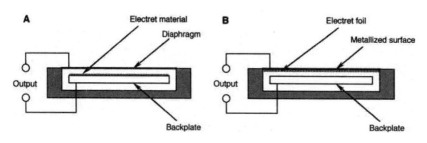

FIGURE 3–18

The electret capsule: electret backplate (A); electret coated diaphragm (B).

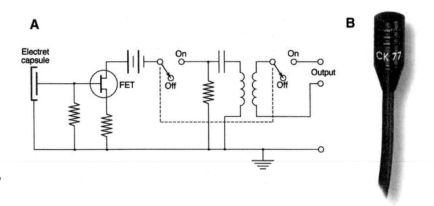

FIGURE 3-19

Complete electrical circuit for an electret microphone (A); photo of a small electret capsule (B). (Photo courtesy of AKG Acoustics.)

INSTRUMENTATION MICROPHONES

Pressure capacitor microphones are generally used for all instrumentation and measurement purposes, and as such they differ slightly in design from those microphones intended for recording and broadcast use. Figure 3–20 presents a cutaway view of a Brüel & Kjaer instrumentation capsule showing a perforated backplate somewhat smaller in diameter than the diaphragm, exposing a back air cavity normally not present in microphones designed for the studio.

Instrumentation microphones are used in the loudspeaker industry, architectural acoustics, and as a primary tools for acoustical response and noise measurements in industrial workplaces.

The most ubiquitous use of these microphones is in the *sound level meter* (SLM), shown in Figure 3–21A. The SLM is the primary instrument for sound survey and noise measurement in our everyday environment. (For routine home sound system setup and level checking, there are a number of low cost sound survey meters, most of them in the $40 range and available from electronics stores.)

The instrumentation microphone is designed for very uniform response, unit to unit, and is provided with accessories to facilitate calibration in the field for variations in temperature and barometric pressure. Polarizing voltages are normally in the 200 V range, which allows somewhat greater separation between diaphragm and backplate, ensuring lower distortion at higher operating levels.

Instrumentation microphones are available in many sizes to cover the frequency range from well below 1 Hz to 200 kHz and beyond, depending on the application. There are models that respond to seismic air movements and have a self-noise floor of -2 dB(A); other models are designed to work at sound pressure levels in the range of 180 dB L_p and above. Figure 3–21B shows the sound pressure level operating ranges of several studio and instrumentation microphones.

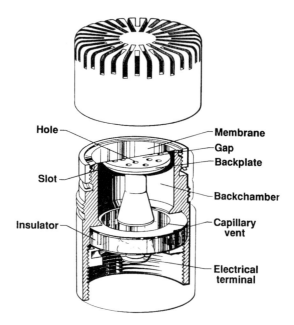

Hole

Membrane

Gap

Backplate

Slot

Backchamber

Insulator

Capillary vent

Electrical terminal

FIGURE 3-20

Cutaway view of an instrumentation microphone. (Figure courtesy of Brüel and Kjaer, 1977.)

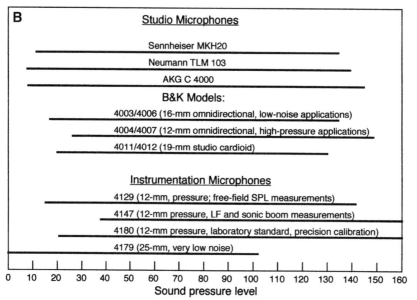

B

Studio Microphones

Sennheiser MKH20

Neumann TLM 103

AKG C 4000

B&K Models:

4003/4006 (16-mm omnidirectional, low-noise applications)

4004/4007 (12-mm omnidirectional, high-pressure applications)

4011/4012 (19-mm studio cardioid)

Instrumentation Microphones

4129 (12-mm, pressure; free-field SPL measurements)

4147 (12-mm pressure, LF and sonic boom measurements)

4180 (12-mm pressure, laboratory standard, precision calibration)

4179 (25-mm, very low noise)

0 10 20 30 40 50 60 70 80 90 100 110 120 130 140 150 160
Sound pressure level

FIGURE 3-21

A modern digital sound level meter (A); operating ranges of selected studio and instrumentation microphones (B). (Photo courtesy of Brüel and Kjaer, 1977.)

SENSITIVITY BANDWIDTH, AND DIAPHRAGM DISPLACEMENT OF CAPACITOR MICROPHONES

Kinsler et al. (1982) present the following simplified equation for the base sensitivity of a capacitor capsule:

$$M = \frac{E_0 a^2}{h 8 T_0} \text{ volt per pascal} \qquad (3.7)$$

where

E_0 = polarizing voltage (V)

h = nominal distance from diaphragm to backplate (m)

a = radius of diaphragm (m)

T_0 = effective diaphragm tension (N/m)

Using the following constants for a 25 mm (1 in) diameter instrumentation microphone:

$a = 8.9 \times 10^{-3}$ m

$T_0 = 2000$ N/m

$E_0 = 200$ volts

$h = 25 \times 10^{-6}$ m

σ_m (surface mass density of diaphragm) = 0.0445 kg/m^2

we can calculate approximate values of microphone sensitivity and upper frequency limit:

$$M = \frac{(200)(8.9)^2(10^{-6})}{25(10^{-6})8(2000)} = 0.04 \text{ V/Pa}$$

Using a more accurate equation, Zukerwar (1994) arrived at a closer value of 0.03 V/Pa. The simplified equation gives us only a reasonable approximation.

Moving on to an equation for the upper frequency limit (f_H) of the microphone, we have (Wong and Embleton, 1994):

$$f_H = \frac{2.4}{2\pi a} \sqrt{\frac{T_0}{\sigma_m}}$$

$$f_H = \frac{2.4}{6.28(8.9 \times 10^{-3})} \sqrt{\frac{2000}{0.0445}} = 9150 \text{ Hz} \qquad (3.8)$$

Above f_H, the response of the microphone is governed by the degree of damping at the resonance frequency and diffraction effects. The response on axis can extend fairly smoothly up to about 20 kHz.

The LF bandwidth extends to very low frequencies, limited only by the RC (resistance-capacitance) time constant established by the low capacitance of the pressure element and the high value of the series resistance. For the 25 mm diameter microphone discussed here, these values are $C = 15 \times 10^{-12}$ farads and $R = 10^9\, \Omega$. Calculating the rolloff frequency, f_L:

$$f_L = 1/(2\pi RC) = 1/6.28(10^9 \times 15 \times 10^{-12})$$
$$f_L = 1/(6.28 \times 15 \times 10^{-3}) = 10.6\,\text{Hz}$$

DIAPHRAGM DISPLACEMENT

The diaphragm rms displacement in meters is given by:

$$\text{Displacement} = \frac{pa^2}{800T_0} \qquad (3.9)$$

For the 25 mm diameter diaphragm under discussion here, the rms displacement is:

$$\text{Displacement} = (8.9 \times 10^{-3})^2 = 5 \times 10^{-11}\,\text{m}$$

It is evident that the diaphragm displacement in the instrumentation microphone discussed here is considerably less than in the studio microphone discussed above.

THE DYNAMIC PRESSURE MICROPHONE

The dynamic microphone is based on principles of electricity and magnetics dating from the 19th century. Because of their relatively low output, it was not until the coming of electrical amplification that these microphones found their place in commerce and early radio broadcasting. Today, we find relatively few dynamic pressure microphones in widespread use, their role having been taken over for the most part by the electret.

The dynamic microphone is also referred to as the *electrodynamic*, *electromagnetic*, or *moving coil* microphone. It is based on the principle of magnetic induction in which a conductor, or wire, moving across a magnetic field has induced along it a voltage proportional to the strength of the magnetic field, the velocity of the motion, and the length of the conductor crossing the magnetic field. The governing equation is:

$$e(t) = Blu(t) \qquad (3.10)$$

where $e(t)$ is the instantaneous output voltage (V), B is the magnetic flux density (T), l is the length of the conductor (m), and $u(t)$ is the instantaneous velocity of the conductor (m/s). Since B and l are constant, the output voltage is directly proportional to the instantaneous velocity of the conductor.

The basic principle of magnetic induction is shown in Figure 3–22, which shows the vector relationship among flux density, conductor orientation, and conductor velocity. In a microphone, the normal

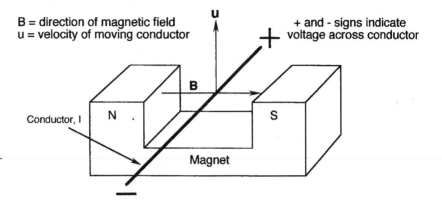

FIGURE 3–22 ————

Basic principle of magnetic induction.

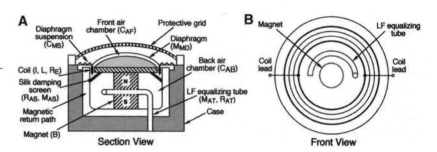

FIGURE 3–23 ————

Section view (A) and front view (B) of a moving coil microphone assembly (protection grid removed in front view).

realization is in the form of a multi-turn coil of wire placed in a radial magnetic field. Section and front views of a typical dynamic pressure microphone are shown in Figure 3–23. In this form, the action is the reciprocal of the modern dynamic loudspeaker.

In a plane progressive sound field, air particle velocity and pressure are in phase. Therefore, for flat electrical output across the conductor, the conductor must execute constant velocity across the entire frequency band in a constant pressure sound field. Since the coil/diaphragm assembly consists of mechanical mass and compliance, it will exhibit mechanical resonance; and this is normally designed to be near the *geometric mean* of the intended frequency response of the microphone. The geometric mean between two quantities along the same numerical scale is defined as:

$$\text{Geometric mean} = \sqrt{\text{lower quantity} \times \text{higher quantity}} \quad (3.11)$$

For a typical response extending from 40 Hz to about 16 kHz, the design resonance is:

$$f_0 = \sqrt{40 \times 16{,}000} = 800 \text{ Hz}$$

The microphone is thus designed to be resistance controlled over its useful frequency range through the application of external damping of the diaphragm's motion.

Figure 3–24 shows the response of an undamped diaphragm (curve 1), along with the effects of increased damping (curves 2 through 5). Note

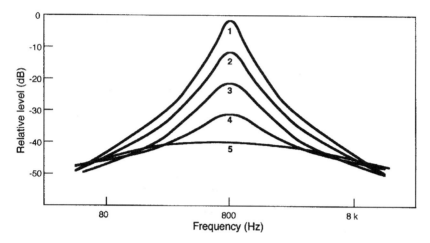

FIGURE 3-24

Basic diaphragm response control, the effect of externally damping the diaphragm.

that, as the damping increases, the midband output becomes flatter, but at the expense of midband sensitivity. To achieve satisfactory frequency response while maintaining a useful overall sensitivity, several design techniques are employed. The diaphragm is damped by placing layers of silk or thin felt in the region of the gap that separates the coil from the back of the air chamber. In general, the MF resonance peak is reduced by about 25–35 dB, producing the response shown in curve 5. To add more midrange damping in this manner would result in unsatisfactory output sensitivity.

The next step is to address the LF response falloff, and this is accomplished by inserting a long, narrow tube into the back air chamber exiting to the outside air. The tube dimensions are chosen so that the air mass in the tube will resonate with the compliance provided by the internal air chamber itself. This Helmholz resonance is chosen to be at some frequency in the 40–100 Hz range. A final step is to compensate for the HF falloff, and this is done by creating a small resonant chamber just inside the diaphragm, tuned in the range 8–12 kHz. These added LF and HF resonances boost the microphone's output in their respective frequency regions.

The design process of damping the primary diaphragm resonance, while adding both LF and HF peaks in the response can best be seen through analysis of the equivalent circuit (after Beranek, 1954), as shown in Figure 3–25.

At LF the response is largely determined by the resonance set up by the bass equalization tube and the back air chamber, and the equivalent circuit in this frequency range is shown in Figure 3–26A.

At MF the response is governed by the damping provided by the silk layers just below the coil, and the equivalent circuit in this frequency range is shown in Figure 3–26B.

At HF the response is determined largely by the resonance set up by the mass of the diaphragm and the compliance of the front air chamber

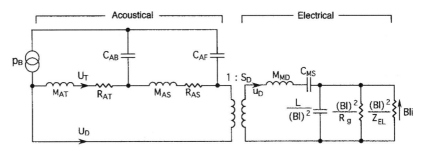

Where:

p_B = rms pressure at front of microphone
M_{AT} = acoustic mass of air in LF equalization tube
R_{AT} = acoustic resistance of air in LF equalization tube
U_D = rms air volume velocity at diaphragm
U_T = rms air volume velocity in LF equalization tube
C_{AB} = compliance of back air chamber
M_{AS} = acoustic mass of screen behind diaphragm
R_{AS} = acoustic resistance of screen behind diaphragm
C_{AF} = compliance of front air chamber
M_{MD} = mass of diaphragm
C_{MS} = compliance of diaphragm
L = coil inductance (henry)
R_E = dc resistance of coil, ohm
Z_{EL} = electrical impedance of coil
B = flux density of magnetic gap (tesla)
l = length of wire in coil (m)

FIGURE 3–25

Simplified equivalent circuit of the dynamic pressure microphone. (Data after Beranek, 1954.)

directly below it. The equivalent circuit in the frequency range is shown in Figure 3–26C, and the net overall response is shown in Figure 3–26D.

We can see that there is much design leeway here, and some pressure dynamic microphones have been designed with more resonant chambers than we have shown here. With care in manufacturing, response tolerances for dynamic pressure microphones can be maintained within a range of ±2.5 dB from 50 Hz to about 15 kHz, as measured on axis. We can appreciate that the design of a good dynamic pressure microphone results from a combination of physics, ingenuity, and years of design practice. Such questions as where to place damping materials and where to assign the LF and HF resonances require experience, and the many variables provide for tailored response for a given microphone application. See Souther (1953) for a practical approach to the design of a dynamic pressure microphone for general application. See also Beranek (1954) and Kinsler et al. (1982).

TYPICAL DIMENSIONS AND PHYSICAL QUANTITIES

The microphone diaphragm may be made of duralumin, a stiff, lightweight alloy of aluminum, or one of many stable plastic materials that can be molded to tight tolerances in thin cross-section. The dome normally moves as a unit, with mechanical compliance provided by reticulated treatment in the outer detail of the dome. Small neodymium magnets are very common in modern designs. Typical physical values for

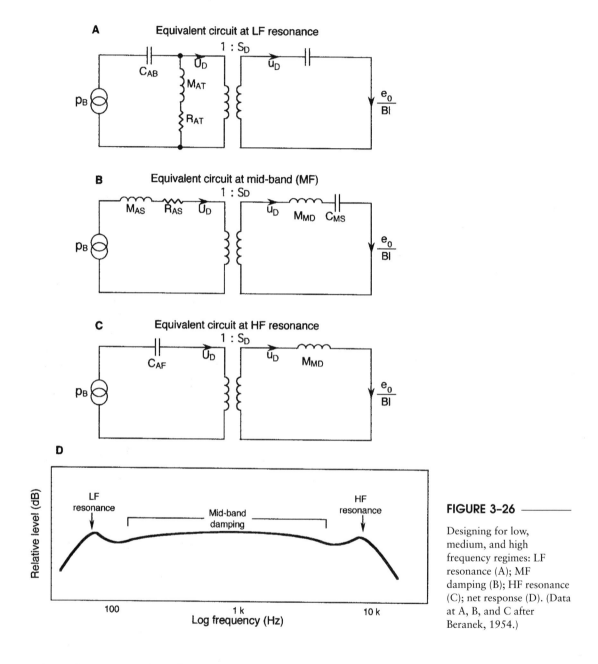

A Equivalent circuit at LF resonance

B Equivalent circuit at mid-band (MF)

C Equivalent circuit at HF resonance

D

FIGURE 3-26 ————

Designing for low, medium, and high frequency regimes: LF resonance (A); MF damping (B); HF resonance (C); net response (D). (Data at A, B, and C after Beranek, 1954.)

a dynamic pressure microphone are:

B = 1.5 T (tesla)

l (length of wire in gap) = 10 m

Diaphragm/coil radius = 9 mm

Diaphragm displacement for 94 dB L_P at 1 kHz = 2×10^{-2} μm rms

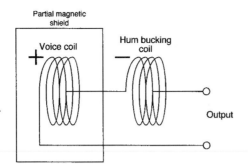

FIGURE 3-27 ————

Electrical circuit showing hum-bucking coil in a dynamic microphone.

Mass of moving system = 0.6 grams

Electrical impedance = 150–125 Ω

Sensitivity range = 1.5–2.5 mV/Pa

Many dynamic microphones, both omnidirectional and directional, incorporate a "hum-bucking" coil to cancel the effect of stray power line induction at 50/60 Hz. This is a low-resistance coil in series with the voice coil and located externally to the magnet structure of the microphone. It is wound reversely and located along the same axis as the voice coil. It is designed to produce the same induced signal that the voice coil would pick up in the stray alternating magnetic field; thus the two induced signals will cancel. The hum-bucking coil is shown in the schematic of Figure 3–27.

OTHER PRESSURE MICROPHONE TRANSDUCERS

Numerous other transducing principles have been used in the design of pressure microphones, including the following:

- *Loose contact principle.* These methods were discussed in Chapter 1 since they were key items in the early history of microphone technology. Only the loose carbon particle microphone has remained a viable design in telephony over the years, and those readers interested in its ongoing development are referred to the Bell Telephone Laboratories (1975) book, *History of Engineering and Science in the Bell System.*

- *Magnetic armature techniques.* This technology was described in Chapter 1 as an essential part of Bell's original telephone patent, but the microphone application is not current today. Olson (1957) discusses it in detail. The system is bilateral and can be employed as either a microphone or a loudspeaker.

- *Electronic techniques.* Olson (1957) describes a microphone in which an electrode in a vacuum tube is mechanically actuated by an external diaphragm, thus varying the gain of the vacuum tube. The technique is entirely experimental and has not been commercialized.

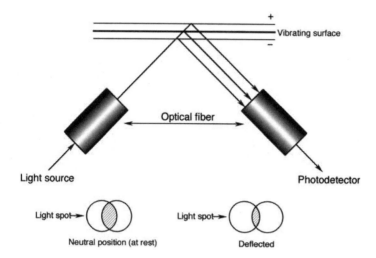

FIGURE 3-28

Details of an optical microphone. (Data after Sennheiser.)

- *Magnetostriction.* Certain magnetic materials undergo a change in dimension along one axis under the influence of a magnetic field. The technique has applications in mechanical position sensing.
- *Optical techniques.* Sennheiser has demonstrated an optical microphone in which a light beam shines on a diaphragm with a modulated beam reflecting onto a photodiode. Details are shown in Figure 3–28. The device can be made quite small and has no electronic components at the diaphragm (US Patent 5,771,091 dated 1998).
- *Thermal techniques.* Olson (1957) describes a microphone consisting of a fine wire heated by direct current. Air currents due to sound propagation around the wire change its resistance proportionally. If the wire can be biased by direct current air flow, then minute changes in wire resistance due to sound pressure in the direction of the air flow can be detected. The technique is very promising in the area of acoustical intensity measurement.

PIEZOELECTRIC MICROPHONES

During the 1930s to the 1960s, the piezoelectric, or crystal, microphone was a significant factor in the market for home recording and small-scale paging and sound reinforcement activities. Today, low-cost dynamic and electret microphones have all but displaced the piezoelectric microphone. (The term *piezo* comes from the Greek *piezen*, meaning pressure.)

Certain crystalline materials, such as Rochelle salts (potassium sodium tartrate), ADP (ammonium dihydrogen phosphate), and lithium sulfite, exhibit the property of developing a voltage along opposite sides when they are flexed or otherwise subjected to deformation. The effect is bilateral, and the elements can also be used as HF loudspeakers.

FIGURE 3-29

Details of a piezoelectric bimorph element (A); section view of a piezoelectric microphone (B).

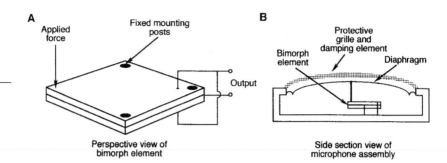

A
Applied force
Fixed mounting posts
Output
Perspective view of bimorph element

B
Bimorph element
Protective grille and damping element
Diaphragm
Side section view of microphone assembly

The crystal structures must be cut and arrayed along the appropriate crystalline axes to produce the desired output voltage. Most piezoelectric microphones are assembled from so-called *bimorph* piezo structures in which adjacent crystal elements are cemented to each other in an opposite sense so that a push-pull output result. Figure 3–29A shows a typical bimorph structure, consisting of two crystal elements cemented together with metal foil on each conducting surface. The element is secured on three corners, and the free corner is driven by the diaphragm through a connecting member.

Figure 3–29B shows a section view of a typical crystal microphone. The diaphragm motion is connected to the free corner of the bimorph, and the output voltage is fed downstream to a preamplifier. Since the output voltage from the bimorph is proportional to signal displacement, the diaphragm is normally tuned to a high frequency, with external mechanical damping, as shown in the figure.

Typical crystal microphones exhibit output sensitivities in the midband of about 10 mV/Pa and, as such, can be used to drive high-impedance preamplifiers directly. The cable length between microphone and preamplifier is limited because of shunt capacitance HF losses in the cable.

In the areas of hydrophones (sensors used for underwater sonar signal transmission and detection), mechanical positioning, and very high acoustical level measurements, piezo electric elements are useful, in that their mechanical impedance characteristics are better suited for these applications than for normal audio use.

Those readers wanting more information on piezoelectric microphones are referred to Beranek (1954) and Robertson (1963).

C H A P T E R   4

THE PRESSURE GRADIENT MICROPHONE

INTRODUCTION

The *pressure gradient* microphone, also known as the *velocity* microphone, senses sound pressure at two very closely spaced points corresponding to the front and back of the diaphragm. The diaphragm is thus driven by the *difference*, or *gradient*, between the two pressures. Its most common form, the dynamic ribbon (diaphragm) microphone, was developed to a high degree of technical and commercial success by Harry Olson of RCA during the 1930s and 1940s and dominated the fields of broadcasting and recording in the US through the mid-1950s. The BBC Engineering Division in the United Kingdom was also responsible for significant development of the ribbon, extending its useful response to 15 kHz at the advent of FM radio transmission in the mid-1950s.

The use of the term *velocity* derives from the fact that the pressure gradient, at least at long wavelengths, is virtually proportional to air particle velocity in the vicinity of the diaphragm or ribbon. However, the more common term is pressure gradient, or simply gradient, microphone. The basic pressure gradient microphone has a "figure-8"-shaped pickup pattern and its pickup pattern is often referred to as *bidirectional*.

DEFINITION AND DESCRIPTION OF THE PRESSURE GRADIENT

By way of review, the basic action of the pressure microphone is shown in Figure 4-1A and B. The mechanical view shown at A is equivalent to the physical circuit shown at B. The significance of the plus sign in the circle is that a positive pressure at the microphone will produce a positive voltage at its output. The circle itself indicates that the microphone is equally sensitive in all directions.

A mechanical view of the gradient microphone is shown at C. It is open on both sides of the diaphragm and is symmetrical in form. The equivalent physical circuit is shown at D. Here, there are two circles, one positive and the other negative, which are separated by the distance *d*. The significance of the negative sign is that sound pressure entering the back side of the microphone will be in polarity opposition to that entering from the front. The distance *d* represents the equivalent spacing around the microphone from front to back access to the diaphragm.

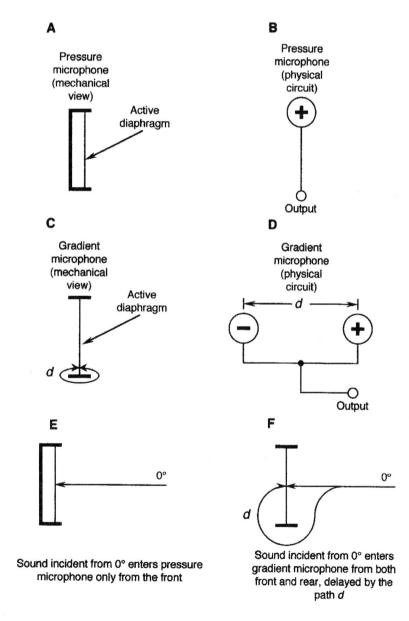

FIGURE 4-1

Definition of the pressure gradient. Mechanical view of pressure microphone (A); physical circuit of pressure microphone (B); mechanical view of pressure gradient microphone (C); physical circuit of pressure gradient microphone (D); 0° sound incidence on pressure microphone (E); sound incidence on pressure gradient microphone (F).

A

Pressure microphone (mechanical view)

Active diaphragm

B

Pressure microphone (physical circuit)

Output

C

Gradient microphone (mechanical view)

Active diaphragm

d

D

Gradient microphone (physical circuit)

d

Output

E

0°

Sound incident from 0° enters pressure microphone only from the front

F

0°

d

Sound incident from 0° enters gradient microphone from both front and rear, delayed by the path *d*

Sound incident on the pressure microphone along the 0° axis will enter only from the front, as shown at E, while sound incident on the gradient microphone along the 0° axis will enter by the front opening as well as through the back opening delayed by the path d. This is shown at F.

The gradient microphone has the capability of determining the propagation angle of a wave. This can be seen in the upper part of Figure 4–2A. A portion of a plane wave is shown along with locations of the two microphone openings placed in line with the wave propagation. These two points relate directly to the points indicated on the sine wave shown below and represent the two openings of the gradient microphone. Now, if the two openings are positioned as shown in Figure 4–2B, there will be no pressure gradient between them and thus no signal at the microphone's output.

The pressure gradient increases with frequency, as shown in Figure 4–3. Here, we show the gradients that exist at LF, MF, and HF when the pressure sensing points are in line with wave propagation. For long wavelengths, the gradient is directly proportional to frequency and is of the form:

$$ge^{j\omega t} = K_1 \, j\omega p e^{j\omega t} \tag{4.1}$$

where $ge^{j\omega t}$ represents the instantaneous value of the gradient, K_1 is an arbitrary constant, and $pe^{j\omega t}$ is the instantaneous value of pressure. If the air particle velocity is uniform with respect to frequency, the $K_1 j\omega$ multiplier in the right side of equation (4.1) indicates that the value of the pressure gradient will be proportional to frequency, rising at a rate of 6 dB/octave, and that the phase of the pressure gradient will be advanced 90° with respect to pressure. This can in fact be seen through a study of the graphical data shown in Figure 4–3.

For a pressure gradient microphone to exhibit flat response in a uniform pressure field, there must be some form of equalization present that

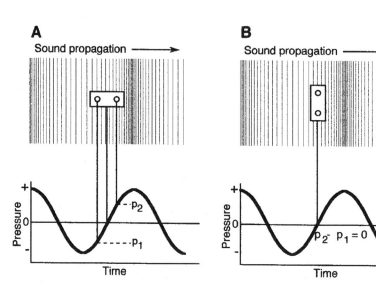

FIGURE 4–2

Pressure gradient measured longitudinally in a progressive plane wave is equal to $p_2 - p_1$ (A); when placed transversely in the progressive plane wave the gradient is zero (B).

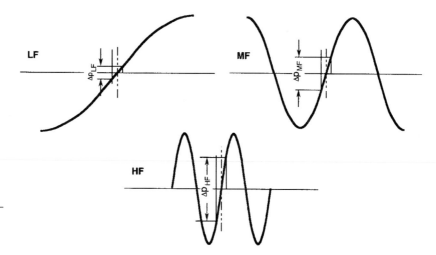

FIGURE 4-3 ——————

Pressure gradient at LF, MF, and HF.

exactly counteracts the HF rise of the pressure gradient. In the case of the dynamic gradient microphone the electrical output is:

$$E\,e^{j\omega t} = B1Ue^{j\omega t} \tag{4.2}$$

where E is the rms output voltage of the microphone, B is the flux density in the magnetic gap (tesla), l is the length of the moving conductor in the magnetic field (m), and U is the rms velocity of the conductor (m/s).

This output voltage would tend to rise 6 dB/octave at HF if the microphone were placed in a sound field that exhibited rising HF air particle velocity. The required equalization can be done electrically, but is accomplished very simply by using a *mass-controlled* ribbon or diaphragm of the form $K_2/j\omega$. The combination of the pressure gradient and the mechanical HF rolloff is then:

$$\text{Net response} = (K_1\,pj\omega e^{j\omega t})/(K_2/j\omega) = (K_1/K_2)pe^{j\omega t} \tag{4.3}$$

which is now proportional to the rms pressure and is flat over the frequency band.

With the capacitor form of the gradient microphone, flat electrical output in a flat sound pressure field depends on constant displacement of the diaphragm over the entire frequency range. Since the driving pressure gradient rises 6 dB/octave at HF, it will produce a constant particle displacement with frequency at the capacitor diaphragm. Therefore, if the capacitor diaphragm's mechanical damping is large compared to its compliance and mass reactance, then the required equalization will be attained.

DIRECTIONAL AND FREQUENCY RESPONSE OF GRADIENT MICROPHONES

Figure 4–4A shows the basic "figure-8" response of the gradient microphone. It is important to note that the response is maximum at

0° and 180°, but that the polarity of the signal is negative (reversed) in the back hemisphere relative to the front hemisphere. Response at ±45° is 0.707, which is equivalent to −3 dB. Response at ±90° is effectively zero. The directional response equation in polar coordinates is:

$$\rho = \text{cosine } \theta \tag{4.4}$$

where ρ represents the magnitude of the response and θ represents the polar angle. Figure 4–4B shows the response plotted in decibels.

The frequency range over which the desired polar response is maintained depends on the size of the microphone and the effective distance between the two sensing points, or distance from the front of the diaphragm around the assembly to the back of the diaphragm. The first null in response takes place when the received frequency has a wavelength that is equal to the distance, d, between the two openings of the microphone.

The pressure gradient, as a function of frequency, is shown in Figure 4–5A. The response rises 6 dB per octave, falling to a null in response at the frequency whose wavelength is equal to d, the path around the microphone. The theoretical equalized microphone output is shown at B. However, due to diffraction effects, the actual response at HF may resemble that shown by the dashed line in the figure.

For extended frequency response, we would like d to be as small as possible. If $d/\lambda = 1/4$, the drop in response will only be 1 dB (Robertson, 1963). Solving this equation for a frequency of 10 kHz gives a value of $d = 8.5$ mm (0.33 in). However advantageous this might be in terms of frequency response, the very short path length would produce a relatively small pressure gradient, and the resultant output sensitivity of the microphone would be low. Fortunately, diffraction effects work to our advantage, maintaining excellent on-axis frequency response even when a substantially longer front-to-back path is used. This will be discussed in the following section.

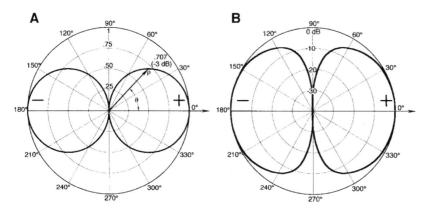

FIGURE 4-4

Directional response in polar coordinates of the gradient microphone; sound pressure gradient response (A); response level in decibels (B).

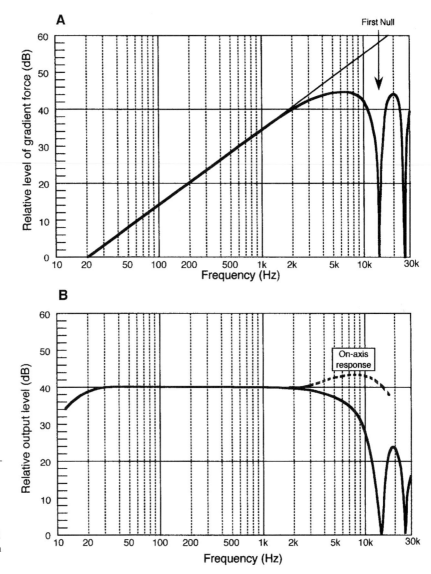

FIGURE 4–5

Pressure gradient force versus frequency, unequalized (A) and equalized (B), for a typical ribbon microphone (typical on-axis HF response shown by dashed curve).

MECHANICAL DETAILS OF THE DYNAMIC PRESSURE GRADIENT MICROPHONE

Mechanical and electrical views of a ribbon gradient microphone are shown in Figure 4–6. The ribbon itself is normally made of corrugated aluminum foil and is about 64 mm (2.5 in) in length and 6.4 mm (0.25 in) in width. It is suspended in the magnetic gap with provisions for adjusting its tension. In newer designs the thickness of the ribbon is often in the range of 0.6 microns (2.5×10^{-5} in), and the length in the range of

25 mm (1 in). The mass of the ribbon, including its associated air load, is about 0.001 to 0.002 grams, and the subsonic resonance frequency of the ribbon is in the range of 10 to 25 Hz. Because of the small gap between the ribbon and polepieces, there may be considerable viscous damping on the ribbon in the region of resonance. Flux density in the gap is normally in the range of 0.5 to 1 T (tesla).

Early ribbon microphones were quite large, with heavy, relatively inefficient magnets. The fragility of the ribbon microphone was well known to all who used them; some ribbons tended to sag slightly over time, and a good puff of wind could actually deform them. Because of their flexibility, ribbons may, like a string, vibrate at harmonic multiples of their fundamental frequency (Robertson, 1963). This is not often encountered in well-damped designs, however.

A closer look at the workings of a ribbon microphone will often reveal a careful use of precisely designed and shaped screens, made of fine metal mesh or of silk, whose purposes are to make fine adjustments in frequency response. As shown in Figure 4–7A, a metal mesh screen has been mounted around the polepiece assembly in order to compensate for LF losses due to the high damping on the ribbon in the region of resonance.

The fine mesh of the structure has been chosen to provide a fairly uniform acoustical impedance over the operating frequency range of the microphone. The vector diagram at B shows its effect at LF. The loss introduced by the screen over the path d results in a lower value of pressure (p_2') at the rear of the ribbon. This actually produces a larger

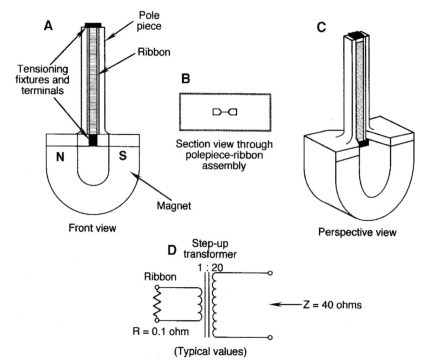

FIGURE 4–6

A ribbon gradient microphone; front view (A); top section view (B); perspective view (C); electrical circuit (typical) associated with the ribbon microphone (D). (Data at B and C after Beranek, 1954.)

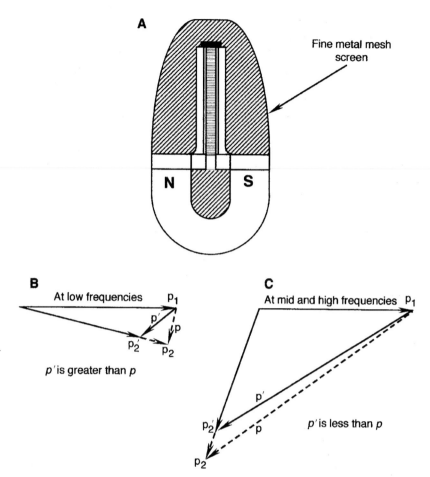

FIGURE 4-7

Correcting LF response in a gradient microphone through acoustical attenuation between front and back. Microphone with screen (A); vector diagram at LF (B); vector diagram at HF (C).

gradient (p') at LF than would exist without the screen (p), and the net effect is a compensating boost at LF.

At MF and HF, the screen's action is as shown at C. Here, the effect of the screen is to cause a slight decrease in the gradient (p'), which may be a small price to pay for correcting the LF response. This design is used to good effect in the Coles model 4038 microphone and is described in detail by Shorter and Harwood (1955).

SENSITIVITY OF THE RIBBON GRADIENT MICROPHONE

There are four main factors affecting the base sensitivity of the ribbon: magnetic flux density in the gap, length and mass of the ribbon, and the length of the front-to-back path. Aluminum foil, since it possesses the best combination of low mass and low electrical resistivity, is perhaps the ideal material for the application. The aspects of internal dimensioning are all interrelated, and any attempt to optimize one will likely result in a compromise elsewhere. For example, doubling the length of the ribbon will raise the sensitivity by 6 dB but will adversely affect vertical directional response at high frequencies. Doubling the path length around the polepieces would also provide a 6-dB increase in sensitivity but would compromise the HF performance of the microphone because of the overall increase in microphone size.

The area where many modern improvements have taken place is in magnetics. The use of higher energy magnets, along with new magnetic circuit materials and topology, has made it possible to achieve enough improvement in base sensitivity to allow further beneficial tradeoffs, including shorter ribbons, with their better behaved polar response and HF performance. In most magnetic designs, the choice and dimensioning of materials is made so that the flux through the polepieces is effectively saturated, and thus uniform throughout the length of the gap.

Because of the extremely low electrical output from the ribbon, a step-up transformer is normally installed in the microphone case. A transformer turns ratio of about 20:1 can be used to match the very low impedance of the ribbon to a 300 ohm load. The open circuit sensitivity of a typical ribbon microphone, operating without an output transformer, is normally in the range of 0.02 mV/Pa. The addition of a step-up transformer raises the system open circuit sensitivity to the range of 0.5–1.0mV/Pa.

RESPONSE CURVES FOR RIBBON GRADIENT MICROPHONES

The design simplicity of the ribbon microphone produces a predictable, uniform pattern up to the frequency where $d = (5/8)\lambda$. Olson (1957) presents the theoretical polar data shown in Figure 4-8. Note that as the wavelength approaches the internal dimensions of the microphone, the pattern flattens slightly on-axis, eventually collapsing to produce zero on-axis response, with polar response of four equal off-axis lobes.

Figure 4–9 shows on-axis response curves for typical commercial ribbon gradient microphones. The curve shown at A is from RCA product literature for the model 44-B ribbon microphone. The data shown at B

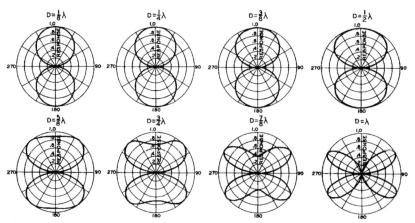

FIGURE 4-8

Polar response of a theoretical ribbon gradient microphone as a function of λ and d. (Olson, 1957.)

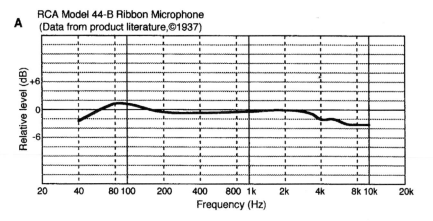

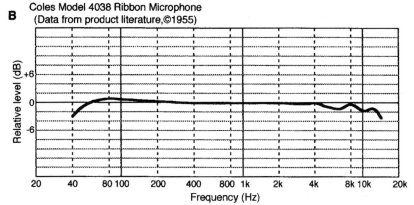

FIGURE 4-9

Typical frequency response curves for ribbon gradient microphones RCA 44-B (A); Coles 4038 (B).

is for the Coles model 4038 microphone. Note that the Coles exhibits slightly flatter and more extended HF response.

Figure 4–10A shows details of the beyerdynamic Model M 130 ribbon microphone. It is perhaps the smallest, and flattest, ribbon microphone available today, with the entire ribbon assembly contained within a sphere 38.5 mm in diameter. Details of the ribbon structure are shown at B, and the microphone's response at 0° and 180° is shown at C.

The more recent AEA model R-84 has much in common with the earlier RCA designs in that it has a large format ribbon. This model is shown in Figure 8–11A, with front and back response curves at B. Note that the vertical scale is expanded and that the response, though extended to 20 kHz, has very much the same general contour as the response curve of the RCA 44.

EQUIVALENT CIRCUIT FOR A RIBBON GRADIENT MICROPHONE

Figure 4–12A shows the equivalent circuit for a ribbon gradient microphone using the impedance analogy (Beranek, 1954). Ignoring the effects of the slits between the ribbon and polepieces and damping in the system, we can simplify the circuit to the form shown at B. Here, the

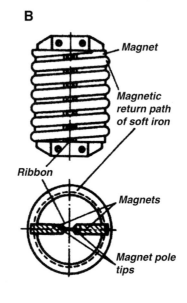

Magnet

Magnetic
return path
of soft iron

Ribbon

Magnets

Magnet pole
tips

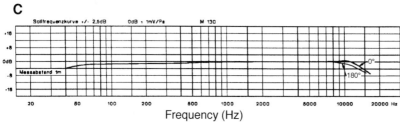

Sollfrequenzkurve +/- 2,5dB   0dB = 1mV/Pa   M 130

Messabstand 1m

0°

180°

Frequency (Hz)

FIGURE 4-10

Beyerdynamic M 130 ribbon microphone: photo (A); details of ribbon structure (B); frequency response (C). (Data courtesy of beyerdynamic.)

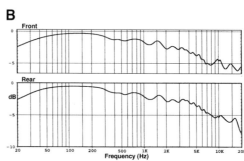

Front

Rear

dB

Frequency (Hz)

FIGURE 4-11

AEA R-84 ribbon microphone: photo (A); front and back response curves (B). (Data courtesy of AEA.)

A   Acoustical circuit (impedance analogy)

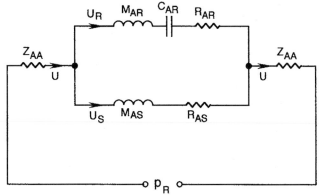

Where:

Z_{AA} = acoustical impedance of medium on one side of diaphragm
U = volume velocity through microphone
U_R = volume velocity of ribbon
U_S = volume velocity of air through slits at edges of ribbon
M_{AR} = acoustic mass of ribbon
C_{AR} = acoustic compliance of ribbon
R_{AR} = acoustic resistance of ribbon
M_{AS} = acoustic mass of slots
R_{AS} = acoustic resistance of slots
p_R = pressure gradient at ribbon

B   Simplified electroacoustical circuit (mobility analogy)

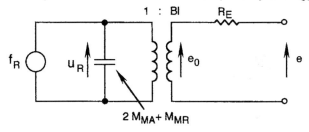

FIGURE 4-12

Equivalent circuits for a ribbon gradient microphone; acoustical circuit (A); simplified electroacoustical circuit (B). (Data after Beranek, 1954.)

Where:

f_R = force on the ribbon
u_R = velocity of ribbon
$2M_{MA} + M_{MR}$ = mass of ribbon and associated air load
e = unloaded electrical output of ribbon
R^0 = dc resistance of ribbon
e_E = unloaded electrical output of microphone system

mobility analogy is used to give an equivalent electromechanical circuit over the range from 50 Hz to about 1 kHz. As can be seen from the circuit, the microphone presents very nearly a pure mass reactance to the acoustical driving signal.

The output voltage is given by:

$$|e_0| = |u|\left(\frac{Bl\rho_0 \Delta d}{2M_{MA}+M_{MR}}\right) S \cos(\theta) \qquad (4.5)$$

where u is the component of air particle velocity perpendicular to the ribbon, Δd is the path length around the microphone, S is the effective area of the ribbon, and $2M_{MA} + M_{MR}$ represents the mass of the ribbon and its associated air load on both sides.

A well-designed ribbon gradient microphone can handle fairly high sound pressure levels in normal studio applications. The moving system is however displacement limited at LF due to the constraints of its design resonance frequency.

THE CAPACITOR GRADIENT MICROPHONE

Figure 4–13 shows three forms of capacitor gradient microphone. The single-ended form shown at A is most common, but its asymmetry does cause a slight HF response difference between front and back. Some designers will place a dummy perforated electrode, which is not polarized, at the front to compensate for this, as shown at B. The push-pull design, shown at C, doubles the output voltage for a given diaphragm excursion relative to the other forms, but is complicated by the required biasing method.

As we discussed in the section entitled Definition and Description of the Pressure Gradient, the capacitor gradient microphone operates in a resistance-controlled mode. This requires that the diaphragm has an undamped resonance in the midband and that considerable viscous air damping be applied to it by the many small holes through the backplate or backplates. There is a practical limit to the amount of damping that is applied, however. Figure 4-14 shows the damping applied progressively, and a typical operating point is indicated at curve 3. Carrying the damping further, in order to gain extended LF response, will result in a very low sensitivity, degrading the self noise floor of the microphone system. Generally, in normal use the LF rolloff will be at least partially compensated by proximity effect (see next section).

Figure 4-15 shows on-axis frequency response curves for a typical high quality capacitor gradient microphone, indicating the degree of LF rolloff that may be tolerated. As in the case of the ribbon microphone, on-axis diffraction effects are used to advantage in maintaining extended HF response.

When judging the merits of various figure-8 microphones, most engineers will choose a ribbon over a capacitor. This may account for the fact that there are relatively few capacitor models available. The natural LF resonance of the ribbon, a vital ingredient in its design, provides extended LF response, which most engineers (and artists) appreciate.

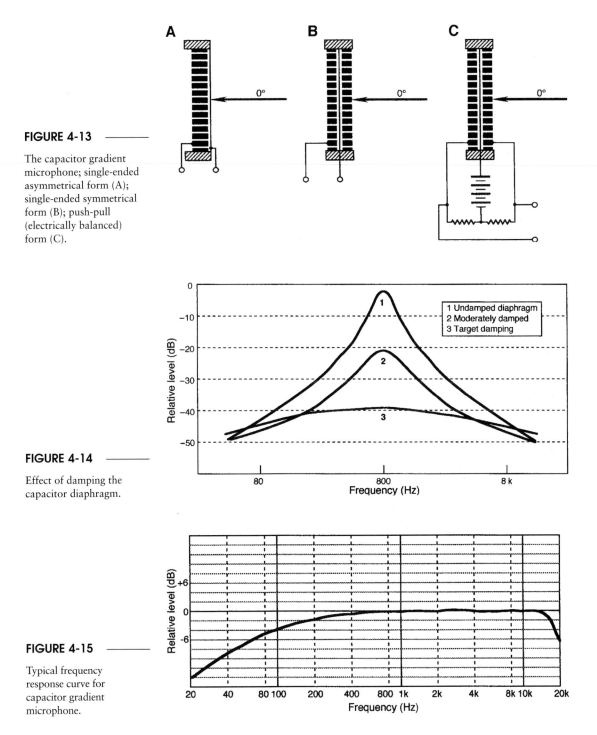

FIGURE 4-13

The capacitor gradient microphone; single-ended asymmetrical form (A); single-ended symmetrical form (B); push-pull (electrically balanced) form (C).

FIGURE 4-14

Effect of damping the capacitor diaphragm.

FIGURE 4-15

Typical frequency response curve for capacitor gradient microphone.

PROXIMITY EFFECT

All directional microphones exhibit *proximity effect*, which causes an increase in LF output from the microphone when it is operated close to a sound source. The effect comes about because, at close working distances, the front and back paths to the diaphragm may differ substantially in their relative distances from the sound source. The gradient component acting on the diaphragm diminishes at lower frequencies, while the inverse square component remains constant with frequency for a given working distance. When this distance is small, as shown in Figure 4-16A, it dominates at LF and causes the output to rise. The data at B shows the relationship between the gradient and inverse square forces on the diaphragm, and the equalized data shown at C indicates the net output of the gradient microphone.

The magnitude of proximity LF boost for a gradient microphone is given by:

$$Boost\ (dB)\ =\ 20\ log\frac{\sqrt{1\ +\ (kr)^2}}{kr} \tag{4.6}$$

where $k = 2p/l$, r is the distance from the sound source (m), and l is the signal wavelength (m). (For example, at an operating distance of 5.4 cm (0.054 m) at 100 Hz, the value of kr is $2p(100)/344 = 1.8$. Evaluating equation (4.6) gives 20.16 dB.)

The rise at LF for several operating distances for a gradient microphone is shown in Figure 4-17. During the 1930s and 1940s, radio crooners soon learned to love the famous old RCA ribbon microphones, with their "well-cushioned" LF response that so often enhanced singers' and announcers' voices.

For symmetrical front and back performance all figure-8 gradient microphones must be *side addressed*, at 90° to the axis of the microphone's

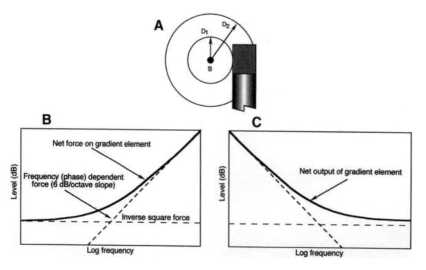

FIGURE 4-16

Proximity effect in the pressure gradient microphone: sound pressure levels proportional to the square of the distance from the source (A); combination of inverse square and frequency-dependent forces on the diaphragm (B); electrical output of microphone system due to the forces shown at B (C).

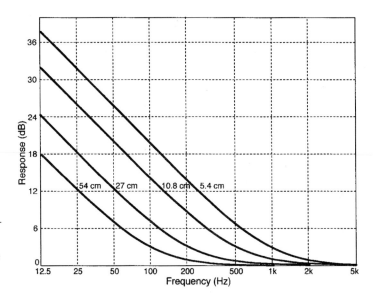

FIGURE 4-17

Typical frequency versus operating distance response curves for a gradient microphone.

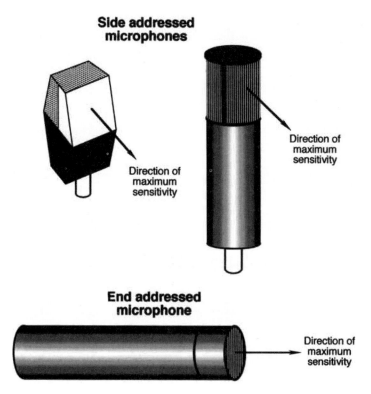

FIGURE 4-18

Axes of maximum sensitivity for a variety of microphone forms.

case, as shown in Figure 4-18. Also in this category would be most studio large diaphragm microphones with adjustable patterns. As a general rule, most small diaphragm models with interchangeable omni and cardioid capsules will be end addressed, as shown in Figure 4-18.

CHAPTER 5

FIRST-ORDER DIRECTIONAL MICROPHONES

INTRODUCTION

The great majority of directional microphones in professional use today are members of the *first-order* cardioid family. The term first-order refers to the polar response equation and its inclusion of a cosine term to the first power, or order. These microphones derive their directional patterns from combinations of the pressure and gradient microphones discussed in the previous two chapters. By comparison, a *second-order* microphone will exhibit a pattern dependent on the *square* of the cosine term. Stated differently, a first-order microphone has response proportional to the pressure gradient, whereas a second-order design has response that is proportional to the gradient of the gradient.

The earliest directional microphones actually combined separate pressure and gradient elements in a single housing, combining their outputs electrically to achieve the desired pattern. Today, most directional microphones with a dedicated pattern have a single transducing element and make use of calculated front-back delay paths to achieve the desired pattern. We will discuss both approaches.

THE FIRST-ORDER CARDIOID FAMILY

The addition of a gradient element and a pressure element is shown in Figure 5–1. In polar coordinates, the omnidirectional pressure element is assigned a value of unity, indicating that its response is uniform in all directions. The gradient element is assigned the value $\cos \theta$, indicating its bidirectional figure-8 pattern:

$$\rho = 0.5 \,(1 + \cos \theta) = 0.5 + 0.5 \cos \theta \qquad (5.1)$$

which is the polar equation for a *cardioid* pattern (so-named for its "heart-like" shape). The geometric construction of the pattern is derived

FIGURE 5-1

Derivation of the cardioid pattern by combination of the pressure (omnidirectional) component and the gradient (bidirectional) component.

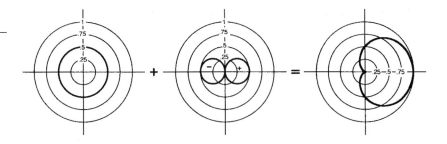

simply by adding the two elements at each bearing angle around the microphone, taking account of the negative back lobe of the gradient component.

Figure 5–2 shows several of the family of polar curves that can be produced by varying the proportions of omnidirectional (pressure) and cosine (gradient) components. Here, the general form of the polar equation is:

$$\rho = A + B \cos \theta \qquad (5.2)$$

where $A + B = 1$.

POLAR GRAPHS OF THE PRIMARY CARDIOID PATTERNS

Figure 5–3 shows graphs of the four main cardioid patterns normally encountered today. The graphs are shown in both linear and logarithmic (decibel) form.

The Subcardioid: Although other A and B ratios may used, this pattern is generally represented by the polar equation:

$$\rho = 0.7 + 0.3 \cos \theta \qquad (5.3)$$

The pattern is shown in Figure 5–3A. The directional response is $-3\,dB$ at angles of $\pm 90°$ and $-10\,dB$ at $180°$. Microphones of this fixed pattern are of relatively recent development, and they have found great favor with classical recording engineers. The subcardioid pattern is sometimes referred to as a "forward oriented omni."

The Cardioid: This pattern is the standard cardioid, and it is represented by the polar equation:

$$\rho = 0.5 + 0.5 \cos \theta \qquad (5.4)$$

This pattern is shown in Figure 5–3B. Directional response is $-6\,dB$ at $\pm 90°$ and effectively zero at $180°$. It is the most widely directional pattern in general use today, its usefulness in the studio deriving principally from its rejection of sound originating at the rear of the microphone.

The Supercardioid: This pattern is represented by the polar equation:

$$\rho = 0.37 + 0.63 \cos \theta \qquad (5.5)$$

The pattern is shown in Figure 5–3C. Directional response is $-12\,dB$ at $\pm 90°$ and $-11.7\,dB$ at $180°$. This pattern exhibits the maximum frontal

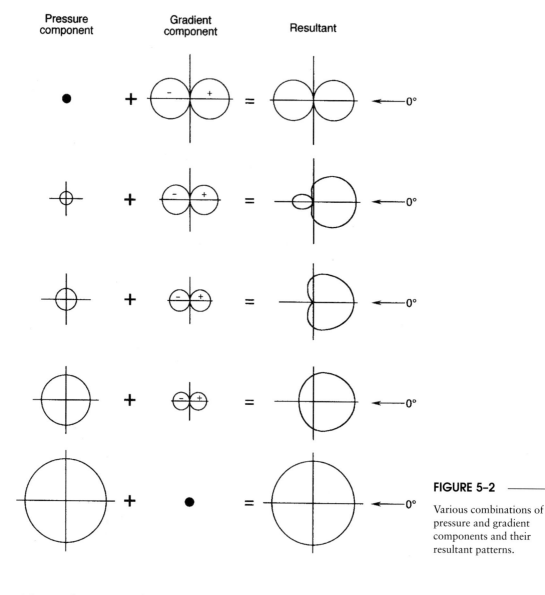

FIGURE 5-2

Various combinations of pressure and gradient components and their resultant patterns.

pickup, relative to total pickup, of the first-order cardioids, and as such is useful for pickup over a wide frontal angle in the studio.

The hypercardioid: This pattern is represented by the polar equation:

$$\rho = 0.25 + 0.75 \cos\theta \qquad (5.6)$$

The pattern is shown in Figure 5–3D. Directional response is $-12\,\mathrm{dB}$ at $\pm 90°$ and $-6\,\mathrm{dB}$ at $180°$. This pattern exhibits the greatest random efficiency, or "reach," in the forward direction of all members of the first-order cardioid family. In the reverberant field, this pattern will provide the greatest rejection, relative to on-axis pickup, of reverberant sound.

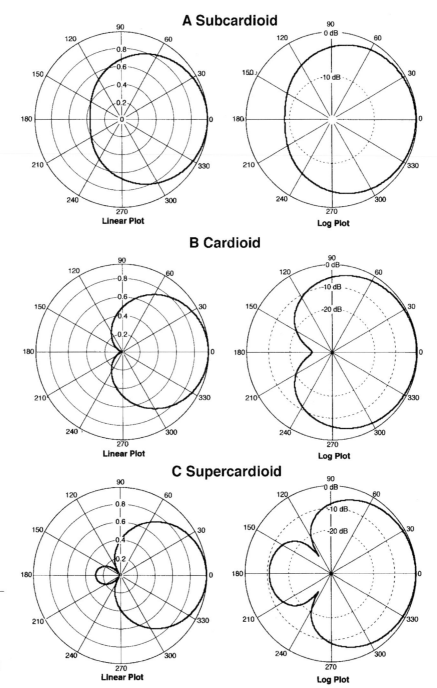

FIGURE 5-3

Polar graphs, linear and logarithmic, for the following patterns: subcardioid (A); cardioid (B); supercardioid (C); and hypercardioid (D).

D Hypercardioid

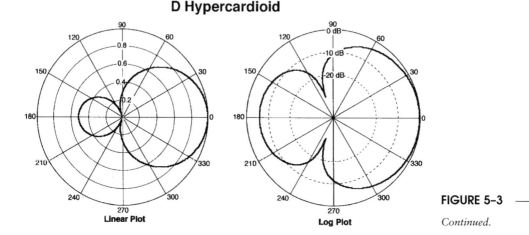

FIGURE 5–3

Continued.

SUMMARY OF FIRST-ORDER CARDIOID PATTERNS

Figure 5–4 shows in chart form the principal characteristics of the first-order cardioids. While most of the descriptions are self evident, two of them need additional comment:

Random Efficiency (RE): RE is a measure of the on-axis directivity of the microphone, relative to its response to sounds originating from all directions. An RE of 0.333, for example, indicates that the microphone will respond to reverberant acoustical power arriving from all directions with one-third the sensitivity of the same acoustical power arriving along the major axis of the microphone. (The term random energy efficiency (REE) is also used.)

Distance Factor (DF): Distance factor is a measure of the "reach" of the microphone in a reverberant environment, relative to an omnidirectional microphone. For example, a microphone with a distance factor of 2 can be placed at *twice* the distance from a sound source in a reverberant environment, relative to the position of an omnidirectional microphone, and exhibit the same ratio of direct-to-reverberant sound pickup as the omnidirectional microphone. These relationships are shown in Figure 5–5 for several first-order patterns.

There are three distinctions in directional patterns that concern the engineer. In many applications, especially in sound reinforcement, the *reach* of the microphone is important for its ability to minimize feedback. In the recording and broadcast studios, the *forward acceptance angle* is important in defining the useful pickup range and how the performers must be arrayed. Under similar operating conditions, the *rejection of off-axis sound* sources may be equally important.

These considerations lead to several important distinctions among subcardioid, cardioid, supercardioid, and hypercardioid patterns. Note for example that the response at ±90° varies progressively from sub- to

SUMMARY OF FIRST-ORDER CARDIOID MICROPHONES

CHARACTERISTIC	PRESSURE COMPONENT	GRADIENT COMPONENT	SUBCARDIOID	CARDIOID	SUPERCARDIOID	HYPERCARDIOID
POLAR RESPONSE PATTERN	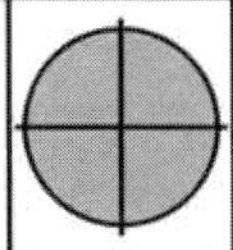	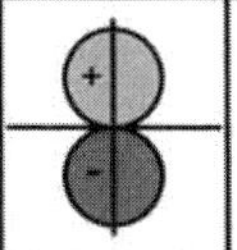	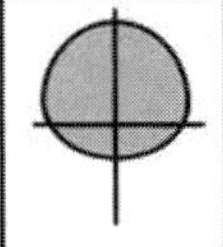	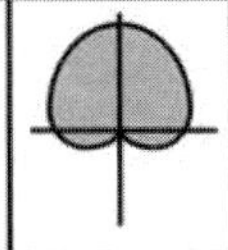	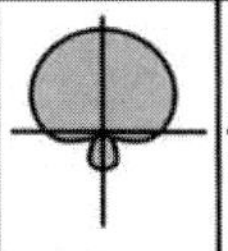	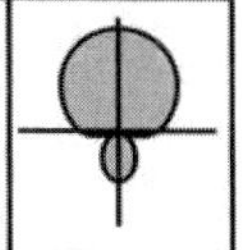
POLAR EQUATION	1	Cos θ	.7 + .3Cos θ	.5 + .5Cos θ	.37 + .63Cos θ	.25 + .75Cos θ
PICKUP ARC 3 dB DOWN	360°	90°	180°	131°	115°	105°
PICKUP ARC 6 dB DOWN	360°	120°	264°	180°	156°	141°
RELATIVE OUTPUT AT 90° (dB)	0	−∞	−3	−6	−8.6	−12
RELATIVE OUTPUT AT 180° (dB)	0	0	−8	−∞	−11.7	−6
ANGLE AT WHICH OUTPUT = ZERO	−	90°	−	180°	126°	110°
RANDOM EFFICIENCY (RE)	1	.333	.55	.333	.268 [1]	.25 [2]
DIRECTIVITY INDEX (DI)	0 dB	4.8 dB	2.5 dB	4.8 dB	5.7 dB	6 dB
DISTANCE FACTOR (DSF)	1	1.7	1.3	1.7	1.9	2

FIGURE 5–4

Characteristics of the family of first-order cardioid microphones.

(1) MAXIMUM FRONT TO TOTAL RANDOM EFFICIENCY FOR A FIRST-ORDER CARDIOID.
(2) MINIMUM RANDOM EFFICIENCY FOR A FIRST-ORDER CARDIOID.

(Data presentation after Shure Inc.)

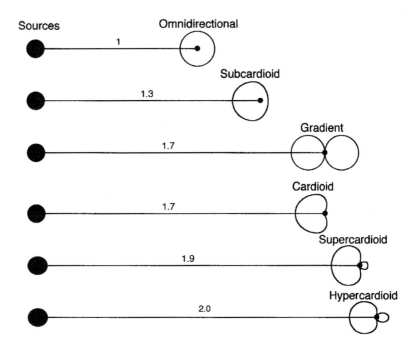

FIGURE 5–5

Illustration of distance factor (DF) for the first-order cardioid family.

hypercardioid over the range of -3 to -12 dB. The effective -6 dB included frontal angle varies progressively from sub- to hypercardioid over the range of 264° to 141°, while the distance factor varies from 1.3 to 2. The back rejection angle varies from 180° for the cardioid to 110° for the hypercardioid. As a function of the B term in equation (5.2), we have the following definitions:

$$\text{Randon Efficiency (RE)} = 1 - 2B + 4B^2/3 \qquad (5.7)$$

$$\text{Front-to-Total Ratio (FTR)} = \text{REF/RE} \qquad (5.8)$$

where

$$\text{REF} = 0.5 + 0.5B + B^2/6 \qquad (5.9)$$

$$\text{Front-to-Back Ratio (FBR)} = \text{REF/REB} \qquad (5.10)$$

where

$$\text{REB} = 0.5 - 1.5B + 7B^2/6 \qquad (5.11)$$

Continuous values of these quantities from $B = 0$ to $B = 1$ are shown in Figure 5–6A, B and C. The figures are labeled to show the maximum values of the functions (Bauer, 1940 and Glover, 1940). It is from this data that the accepted definitions of the supercardioid and hypercardioid patterns were taken.

FIRST-ORDER CARDIOID PATTERNS
IN THREE DIMENSIONS

We can draw microphone patterns only in two-dimensional form, and it is easy to forget that the patterns actually exist in three dimensions, as suggested in Figure 5–7. For microphones that are end addressed, these patterns exhibit consistent rotational symmetry, since the microphone case lies along the 180° angle relative to the major axis. Side addressed microphones may exhibit slightly asymmetrical patterns, inasmuch as the body of the microphone lies along a 90° angle from the major axis.

EXAMPLES OF EARLY DUAL-ELEMENT
DIRECTIONAL MICROPHONES

Figure 5–8 shows details of the Western Electric/Altec 639 multi-pattern microphone. It consists of a ribbon element and a moving coil pressure element in a single housing. The two outputs are summed, and a switch varies the ratio between them producing a family of polar patterns. Because of the relatively large spacing between the elements, accurate frequency response and polar integrity are difficult to maintain beyond about 8 kHz. It can easily be seen that precise alignment of the two elements is critical to their proper summation.

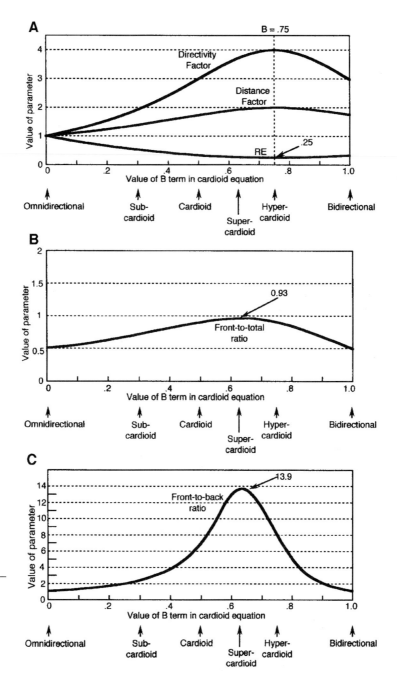

FIGURE 5-6

Graphical illustrations of directivity factor, distance factor, random efficiency, front-to-total ratio, and front-to-back ratio for the first order family.

Working along slightly different lines, Olson (1957) describes a cardioid ribbon microphone, a variant in the 77-DX series, that effectively divided the ribbon into two sections, as shown in Figure 5–9. The upper section of the ribbon provided a gradient operation while the lower section provided pressure operation. Electrical summation of the pressure and gradient effects took place in the ribbon itself.

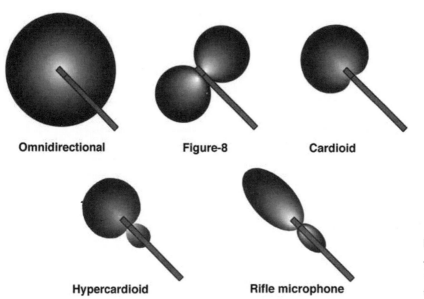

Omnidirectional   **Figure-8**   **Cardioid**

Hypercardioid   **Rifle microphone**

FIGURE 5–7

Three-dimensional
"directivity balloons" for
the first-order family.

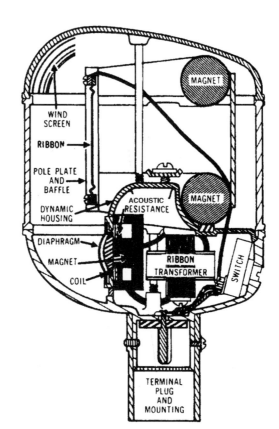

WIND
SCREEN

RIBBON

POLE PLATE
AND
BAFFLE

ACOUSTIC
RESISTANCE

DYNAMIC
HOUSING

MAGNET

MAGNET

DIAPHRAGM

MAGNET

RIBBON
TRANSFORMER

SWITCH

COIL

TERMINAL
PLUG
AND
MOUNTING

FIGURE 5–8

Section view of the
Western Electric model
639 dual-element variable
directional microphone.
(Figure courtesy of Altec.)

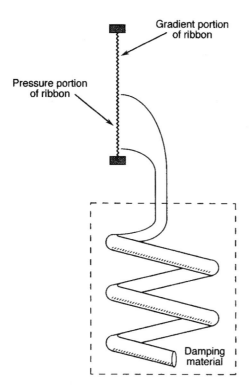

Gradient portion of ribbon

Pressure portion of ribbon

Damping material

FIGURE 5-9

Section view of mechanism of a ribbon cardioid microphone.

PROXIMITY EFFECT IN CARDIOID MICROPHONES

Since the cardioid pattern is a combination of a pressure element (which has no proximity effect) and a gradient element (which does), we would expect the cardioid pattern to exhibit somewhat less proximity effect than the gradient element alone. This is in fact the case, and Figure 5–10 shows the proximity effect on-axis for a cardioid microphone as a function of operating distance. For comparison with the figure-8 pattern, see Figures 4–15 and 4–16.

It is also necessary to consider proximity effect in a cardioid microphone as a function of the operating angle, and this is shown in Figure 5–11 for an operating distance of 0.6 m (24 in). For on-axis operation, the proximity effect will be the same as shown in Figure 5–10 for the same operating distance. Note that at 90° there is no proximity effect. The reason is that at 90° there is no contribution from the gradient element; only the pressure element is operant, and it exhibits no proximity effect. As we progress around the compass to 180° we observe that, for sound sources in the far field, there will be no response. However, as the sound source at 180° moves closer to the microphone, there will be a pronounced increase in proximity effect at very low frequencies. The rapid increase in response with lowering frequency at 180°

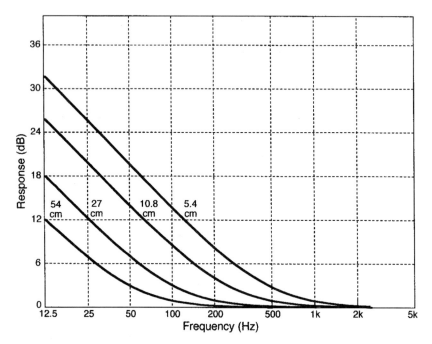

FIGURE 5-10

Proximity effect versus
operating distance for a
cardioid microphone on
axis.

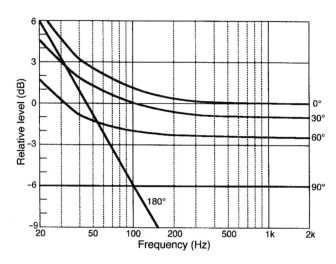

FIGURE 5-11

Proximity effect versus
operating angle at a fixed
distance of 0.6 m (24 in).

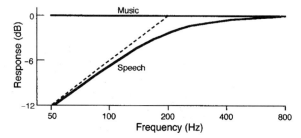

FIGURE 5-12

Typical action of a
high-pass filter to reduce
proximity effect at close
working distances.

is due to the subtractive relationship between gradient and pressure quantities at that angle.

Most directional microphones, dynamic or capacitor, have a built-in switch providing an LF cut in response to partially counteract proximity effect when the microphone is used at short operating distances, as shown in Figure 5–12. This response modification is used primarily for speech applications; for many music applications the proximity LF boost may be considered a benefit.

SINGLE DIAPHRAGM CARDIOID MICROPHONES

A single diaphragm cardioid capacitor microphone is shown in Figure 5–13; representations of sound incident at 0°, 90°, and 180° are shown at A, and the corresponding physical circuits are shown at B. The diaphragm is resistance controlled. Note that there is a side opening in

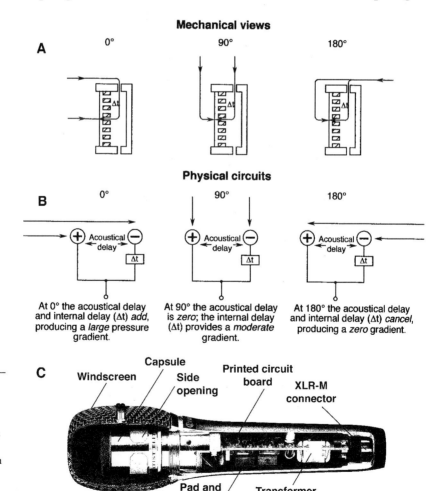

FIGURE 5–13

Basic principle of a single diaphragm cardioid microphone; mechanical views (A); physical circuits (B); cutaway view of a capacitor single diaphragm cardioid microphone (C). (Photo courtesy of Audio-Technica US.)

the microphone case that provides a delay path from the outside to the back of the diaphragm. The net pressure gradient on the diaphragm is the combination of the variable external delay, which is dependent on the angle of sound incidence, and the fixed internal delay. The external acoustical delay and the internal delay have been designed to be equal for on-axis operation.

At 0°, the external acoustical delay and fixed internal delay add to produce a fairly large pressure gradient. This gradient increases with frequency, as we noted in the preceding chapter, and the effect on a resistance-controlled capacitor diaphragm is a constant output over the frequency range. At 90°, the external delay is zero since sound arrives at the front and rear openings of the microphone in-phase. Only the internal delay contributes to the gradient and will be one-half the total delay for 0°. For sound arriving along the 180° axis, the internal and external delays will be equal but in phase opposition, producing a zero gradient and no signal output from the microphone. A cutaway view of a single diaphragm cardioid capacitor microphone is shown at C.

A similar approach can be taken with the dynamic microphone (Bauer, 1941), as shown in Figure 5–14. This basic design is shown at A. With numerous refinements and additions, it has become the basis for today's dynamic "vocal microphones" widely used throughout the music and communication industries. For flat LF response the diaphragm must be mass controlled. This may result in problems of LF handling noise and overload, and most directional dynamic microphones are designed with substantial LF damping, relying on proximity effect at close working distances to equalize the effective frequency response.

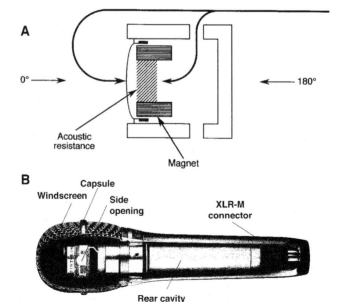

FIGURE 5–14

Basic principle of a single diaphragm dynamic cardioid microphone (A); cutaway view of a typical dynamic single diaphragm cardioid microphone (B). (Photo courtesy of Audio-Technica US.)

As we have seen, the single diaphragm cardioid microphone has much in common with the gradient microphone discussed in the previous chapter. It is in fact a gradient microphone with the addition of a fixed internal delay path behind the diaphragm. Changing the amount of internal delay enables the designer to produce other cardioid types. For example, by shortening the delay appropriately a supercardioid pattern, with its characteristic null at ±126°, and a hypercardioid pattern, with its characteristic null at ±110°, can be attained. A cutaway view of a typical single diaphragm dynamic cardioid microphone is shown at B.

VARIABLE PATTERN SINGLE DIAPHRAGM MICROPHONES

By providing a means for physically changing the internal delay path, variable pattern single diaphragm microphones can be realized. In general, a microphone that offers a selection of patterns forces a trade-off in terms of HF response and pattern integrity as compared with a microphone that has been optimized for a single pickup pattern.

Olson (1957) describes an RCA ribbon microphone in the 77-series that offered a number of pickup patterns. The basic design is shown in Figure 5–15A. A cowl was placed behind the ribbon, and an aperture at the back of the cowl could be varied by a shutter to produce the family of first-order patterns, as shown at B.

The AKG Model C-1000 capacitor microphone is normally operated with a cardioid pattern; however, the grille that encloses the capsule can be removed by the user, and an adapter slipped over the capsule assembly that effectively alters path lengths around the microphone to produce a hypercardioid pattern. This design is shown in Figure 5–16. An external view of the microphone is shown at A; the normal internal configuration is shown at B, and the modified configuration at C.

When the pattern change adapter is placed over the microphone, an additional distributed gradient path is provided. Since it is applied to the positive side of the diaphragm, it works to some degree in opposition to the built-in gradient path on the negative side of the diaphragm. In so doing, it increases the dipole, or figure-8, contribution to the microphone's effective pattern thus producing a hypercardioid pattern.

A very elegant example of a variable pattern, single diaphragm capacitor microphone is the Schoeps MK 6 capsule. Section views of the internal spacing of elements for the three basic patterns are shown in Figure 5–17. The ingenious mechanism fits within a cylindrical housing only 20 mm in diameter. The operation of the microphone is explained as follows:

- *Omnidirectional response.* At A, both inner and outer moving assemblies are in their far left positions, effectively closing off the back openings to the diaphragm and resulting in omnidirectional response.

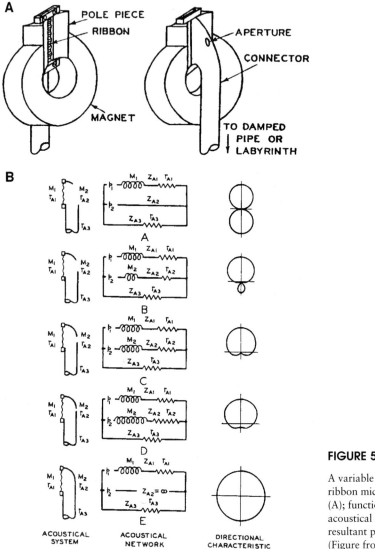

FIGURE 5-15

A variable pattern single ribbon microphone design (A); functional views, acoustical circuits and resultant patterns (B). (Figure from Olson, 1957.)

- *Cardioid response.* At B, the inner moving assembly has been positioned to the right, exposing the rear opening. The delay path through the rear of the microphone is equal to the path around the microphone, so sounds arriving from the rear of the microphone will cancel at the diaphragm.

- *Figure-8 response.* At C, both inner and outer moving assemblies have been positioned to the right, and a new rear opening, symmetrical with the front opening, has been exposed. This front-back symmetry will produce a figure-8 pattern. The left portion of the design is nonfunctional mechanically; essentially, it provides a matching

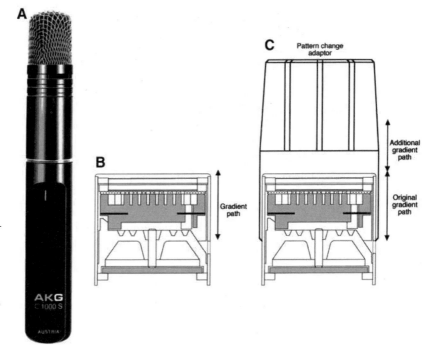

FIGURE 5-16

Details of the AKG model C-1000 cardioid capacitor microphone; external view (A); internal view, normal operation (B); internal view, hypercardioid operation (C). (Figure courtesy of AKG Acoustics.)

acoustical boundary condition on the front side of the microphone which matches that on the back side when the microphone is in the figure-8 mode.

THE BRAUNMÜHL-WEBER DUAL DIAPHRAGM CAPACITOR MICROPHONE

By far the most common variable pattern microphone in use today is the Braunmühl-Weber (1935) dual diaphragm capsule design. The basic design is shown in Figure 5–18A. The backplate is positioned between the two diaphragms; the backplate has shallow holes in each side as well as holes that go through the plate. The purpose of both sets of holes is to provide the required acoustical stiffness and damping on the diaphragms for the various modes of operation of the microphone. An alternate design as used by AKG Acoustics is shown at B; here, there are two perforated backplates separated by a damping element.

As with all capacitor microphones operating as gradient devices, the diaphragms have a midrange tuning and are heavily damped to work in a resistance-controlled mode. The perforations connecting the two diaphragms contribute to a high degree of resistive damping on the diaphragm motion. The air between the two diaphragms acts as a capacitance, and a simplified analogous circuit is shown at C. The challenge in the design is to ensure that the resistive damping is more dominant than

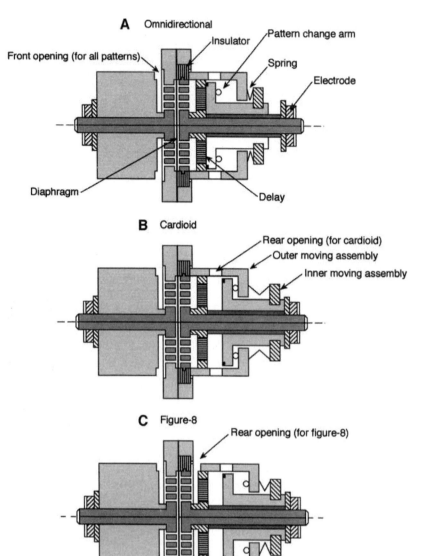

A Omnidirectional

Front opening (for all patterns)
Insulator
Pattern change arm
Spring
Electrode
Diaphragm
Delay

B Cardioid

Rear opening (for cardioid)
Outer moving assembly
Inner moving assembly

C Figure-8

Rear opening (for figure-8)

FIGURE 5-17

Details of the Schoeps
MK 6 single diaphragm
variable pattern capsule.
Omnidirectional response
(A); cardioid response (B);
figure-8 response (C).
(Figure after Schoeps and
Posthorn Recordings.)

the reactive stiffness and mass of the captured air. A photo of a typical dual element capsule is shown at D.

In the example shown in Figure 5–19A, only the left diaphragm has been polarized; the right diaphragm is at zero voltage. First, consider a sound source at a direction of 90° relative to the diaphragms (B). Since both diaphragms are equidistant from the sound source there will be equal and opposite pressure vectors, S_1 and S_2, as shown at B. The pressures will push the diaphragms in and out against the stiffness of the enclosed air.

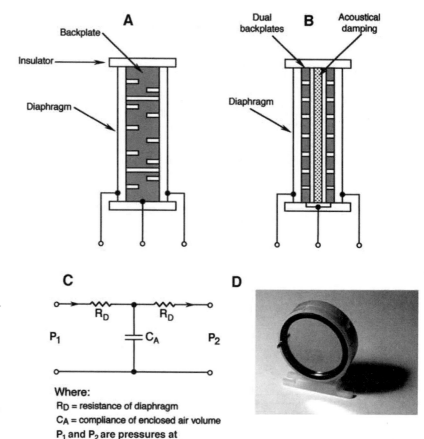

FIGURE 5-18

Section view of the basic Braunmühl-Weber dual diaphragm capacitor microphone (A); an equivalent design used by AKG Acoustics (B); simplified equivalent circuit, impedance analogy (C); view of typical dual backplate capsule (D).

Where:
R_D = resistance of diaphragm
C_A = compliance of enclosed air volume
P_1 and P_2 are pressures at front and back of assembly

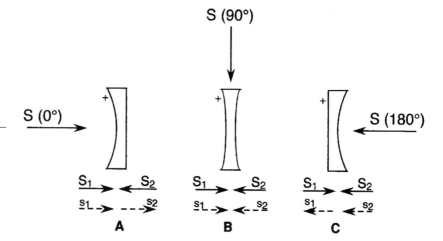

FIGURE 5-19

Vector diagrams showing operation of the dual diaphragm assembly; for sound incident at 0° (A); for sound incident at 90° (B); for sound incident at 180° (C).

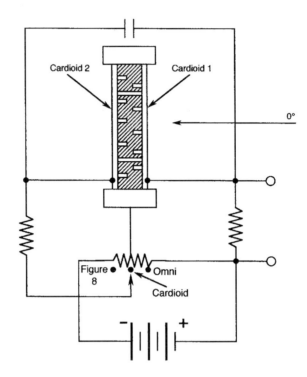

FIGURE 5–20

Electrical circuit for combining the outputs of the dual diaphragms and producing the first-order family of patterns.

For a source at 0°, as shown at A, there will be the same set of vectors plus and additional set of vectors (s_1 and s_2) caused by the pressure gradient effect at the microphone. These pressures will push the diaphragms and enclosed air as a unit because of the interconnection between the two sides of the backplate. The two sets of vectors will combine as shown in the figure, and the two vectors on the back side (180° relative to the signal direction) will, through careful control of damping and stiffness, cancel completely. Only the left diaphragm will move, producing an electrical output. For sounds arriving at 180°, only the right diaphragm will move, as shown at C. Since the back diaphragm is not polarized there will be no electrical output.

In effect, the assembly behaves as two back-to-back cardioid capacitor microphones sharing a common backplate. If both diaphragms are connected as shown in Figure 5–20 and independently polarized, the entire family of first-order patterns can be produced. The data shown in Figure 5–21 illustrates how back-to-back cardioids can be added and subtracted to produce the family of first-order cardioid patterns.

THE ELECTRO-VOICE VARIABLE-D® DYNAMIC MICROPHONE

Wiggins (1954) describes a novel variation on the standard dynamic directional microphone. As discussed above under Proximity Effect in Cardioid Microphones, the standard dynamic directional microphone

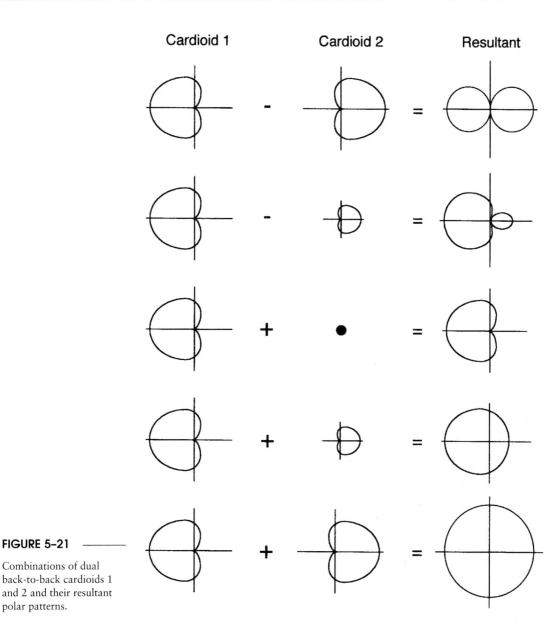

Cardioid 1 **Cardioid 2** **Resultant**

FIGURE 5–21

Combinations of dual
back-to-back cardioids 1
and 2 and their resultant
polar patterns.

relies on a *mass*-controlled diaphragm in order to maintain the extended
LF response. Wiggins designed a dynamic microphone with a *resistance*-
controlled diaphragm and changed the normal single back path into a set
of three back paths to cover LF, MF, and HF actions separately, hence the
term Variable-D (standing for variable distance). The intent of the design
was to produce a wide-band directional dynamic microphone that exhibited better LF response and greater resistance to handling noise and
mechanical shock than is typical of mass-controlled dynamic cardioids.

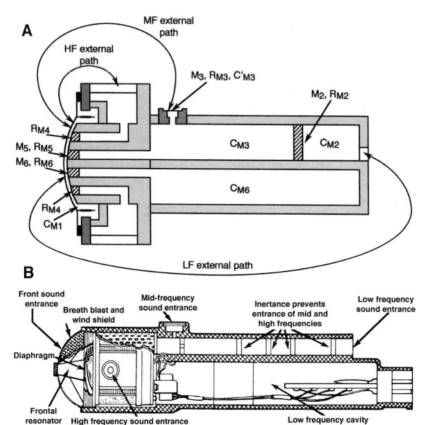

FIGURE 5–22

Schematic view of
Electro-Voice Variable-D
microphone (A); cutaway
view of a typical Variable-D
microphone (B). (Diagram
courtesy of Electro-Voice.)

If a resistance-controlled diaphragm were to be used with a single
back path, the response would roll off at 6 dB/octave at LF. By succes-
sively lengthening the back path for lower frequencies in selective band-
pass regions, the necessary pressure gradient can be maintained, in each
frequency region, to produce flat output.

A schematic view of the microphone is shown in Figure 5–22A, and
the three back path distances (LF, MF, and HF) are clearly indicated. A
cutaway view of a typical Variable-D microphone is shown at B.

An equivalent circuit, shown in Figure 5–23, indicates the complexity of the design in terms of acoustical filtering. The microphone works
by establishing a pressure gradient that is effectively uniform over a large
frequency range (about 200 Hz to 2 kHz). Since the driving force on the
diaphragm is constant over that range, the diaphragm assembly must be
resistance controlled for flat response. At higher frequencies flat on-axis
response is maintained by diffraction effects and careful attention to resonances in the region of the diaphragm. The shunt path in the equivalent
circuit (M_6, R_{M6} and C_{M6}) maintains flat LF response.

Figure 5–24 shows vector diagrams for LF action (A), MF action (B),
and at an intermediate frequency between the two (C). In these diagrams,

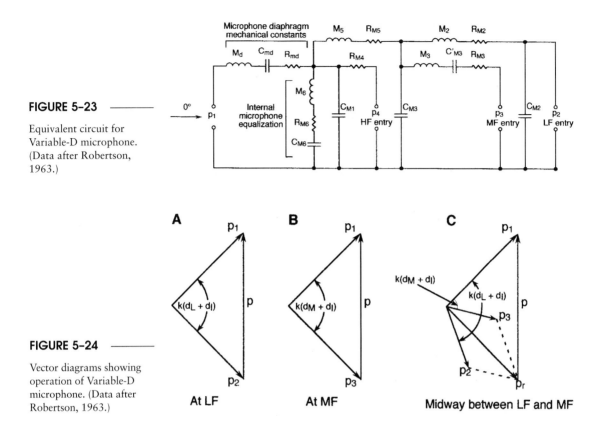

FIGURE 5-23

Equivalent circuit for Variable-D microphone. (Data after Robertson, 1963.)

FIGURE 5-24

Vector diagrams showing operation of Variable-D microphone. (Data after Robertson, 1963.)

k is the wavelength constant, $2p/l$; d_L, d_M, and d_I are, respectively, the LF, MF, and internal delay paths around and through the microphone. The quantities $k(d_L + d_I)$ and $k(d_M + d_I)$ represent, respectively, the phase shift in radians at LF and MF operation. Note that the value of the vector p remains constant, indicating that the gradient is effectively independent of frequency over the range of Variable-D operation.

TWO-WAY MICROPHONES

A persistent problem in the design of single diaphragm directional microphones is the maintenance of target pattern control at very low and very high frequencies. Typical response is shown in Figure 5–25, where it is clear that polar pattern integrity is compromised at both LF and HF. The primary reason for this is the fall-off of the gradient component in the microphone's operation at the frequency extremes, as discussed in Chapter 4 under The Capacitor Gradient Microphone. It is true that more internal damping of the diaphragm would solve this problem, but at a considerable loss of microphone sensitivity. One method for getting around this problem is to design a two-way microphone with one section for LF and the other for HF.

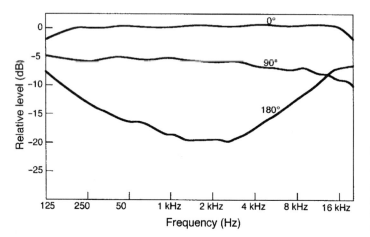

FIGURE 5-25

Typical example of
on- and off-axis frequency
response curves for
single-diaphragm cardioid
microphones.

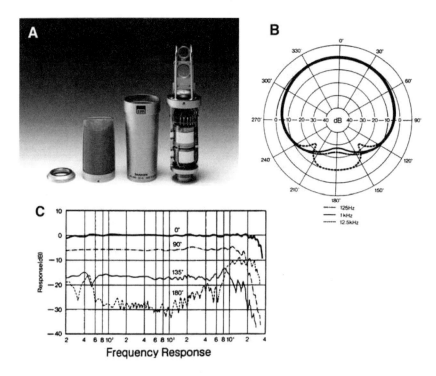

FIGURE 5-26

Details of the Sanken
CU-41 two-way
microphone. Disassembled
view (A); polar response
(B); on- and off-axis
frequency response curves
(C). (Figure courtesy of
Sanken.)

Weingartner (1966) describes such a two-way dynamic cardioid
design. In this design approach, the LF section can be optimized in its
damping to provide the necessary LF pattern control, while the HF section
can be optimized in terms of size and polar performance. It is essential
that the combining networks used to "splice" the two sections together
be carefully designed.

Perhaps the best known capacitor two-way design is the Japanese
Sanken CU-41, which is shown in Figure 5–26. The LF to HF crossover

frequency is 1 kHz, where the wavelength is about 0.3 m (12 in). Since the internal component spacing is small compared to this wavelength, accurate combining can be attained at all normal usage angles with minimum pattern lobing. In optimizing the HF section of this microphone, the ±90° off-axis performance, the cardioid target value of 6 dB, has been maintained out to 12.5 kHz, as can be seen in the family of off-axis curves. Not many cardioid microphones routinely do this. The LF parameters, independently adjusted, result in excellent pattern control to well below 100 Hz.

ADDED FLEXIBILITY IN MICROPHONE PATTERN CONTROL

In its Polarflex system, Schoeps has introduced a method of varying first-order microphone patterns over a wide range. As we saw earlier, omni and figure-8 patterns can be combined to create the entire family of first-order cardioid patterns. Polarflex makes use of separate omni and figure-8 elements and allows the user to combine them as desired over three variable frequency ranges.

As an example, a user can "design" a microphone that is essentially omnidirectional at LF, cardioid at MF, and supercardioid at HF. The transition frequencies between these regimes of operation can be selected by the user. Such a microphone would be useful in recording a large performing group, orchestral or choral, by allowing the engineer to operate at a greater than normal distance while retaining the desired presence

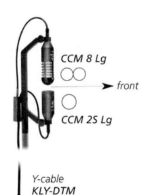

FIGURE 5-27 ————

Photo of Polarflex array showing basic pattern orientations. (Photo courtesy Schoeps.)

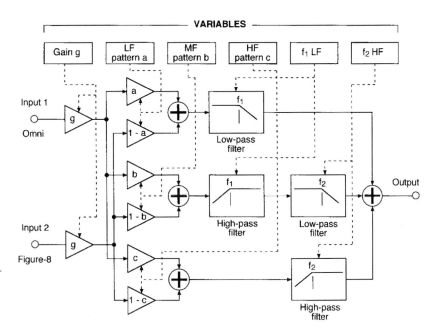

FIGURE 5-28 ————

Polarflex signal flow diagram. (Data after Schoeps.)

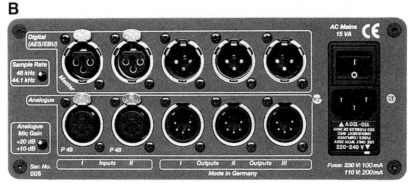

FIGURE 5–29

Front (A) and rear (B) views of DSP-4P processor. (Photos courtesy Schoeps.)

at MF and HF. Another application is in the studio, where the directional characteristics and proximity effect of a vocal microphone can be adjusted for individual applications.

Figure 5–27 shows the basic array. Note that there are both omni and figure-8 elements closely positioned one atop the other. This basic array constitutes in effect a single microphone, and for stereo pickup a pair of such arrays will be required. The elements of the array are processed via the circuit shown in Figure 5–28. The front and rear views of the control module are shown in Figure 5–29.

C H A P T E R 6

HIGH DIRECTIONALITY MICROPHONES

INTRODUCTION

For many applications it is necessary to use a microphone with directional properties exceeding those of the first-order cardioid family. In the film/ video industry, for example, dialog pickup on the shooting set is usually done by way of an overhead boom microphone which must be clearly out of the picture and may be two or more meters away from the actors. Adequate pickup may depend on a highly directional microphone to ensure both speech intelligibility and a subjective sense of intimacy.

Sports events and other activities with high ambient noise levels may require highly directional microphones for noise immunity, and recording in highly reverberant spaces may require such microphones to enhance musical clarity. Field recording of natural events such as bird calls and the like may call for operation at great distances; here, high directionality microphones may be essential in attaining a usable recording.

High directionality microphones generally fall into three categories:

1. Interference-type microphones. These designs achieve high directionality by providing progressive wave interference of high frequency sound arriving off-axis, thus favoring sound arriving on-axis.

2. Focusing of sound by means of reflectors and acoustical lenses. These designs are analogous to optical methods familiar to us all.

3. Second and higher-order designs. These microphones make use of multiple gradient elements to produce high directionality.

INTERFERENCE-TYPE MICROPHONES

Olson (1957) describes a microphone consisting of a number of clustered parallel tubes that differ in length by a fixed amount, as shown in

Figure 6–1A. A microphone transducer is located at the far end, where all sound pickup is combined. For sound waves arriving at 0° incidence, it is clear that the signal at the transducer will be, to a first approximation, the in-phase sum of all contributions. For sound waves arriving from some arbitrary off-axis angle θ, the sum will exhibit a degree of phase cancellation of the contributions due to the differences in the individual path lengths. Olson derives an equation that gives the net sum at the transducer in terms of the received signal wavelength (λ), the overall length of the array (l), and the angle of sound incidence (θ):

$$R_\theta = \frac{\sin \frac{\pi}{\lambda}(l - l \cos \theta)}{\frac{\pi}{\lambda}(l - l \cos \theta)} \qquad (6.1)$$

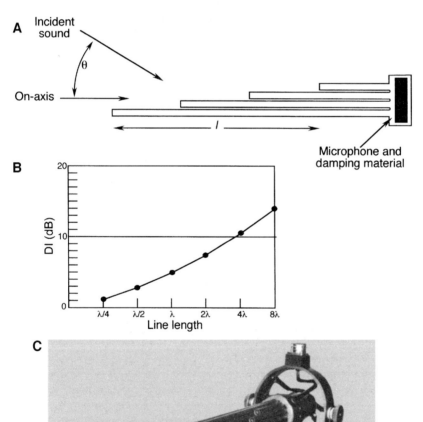

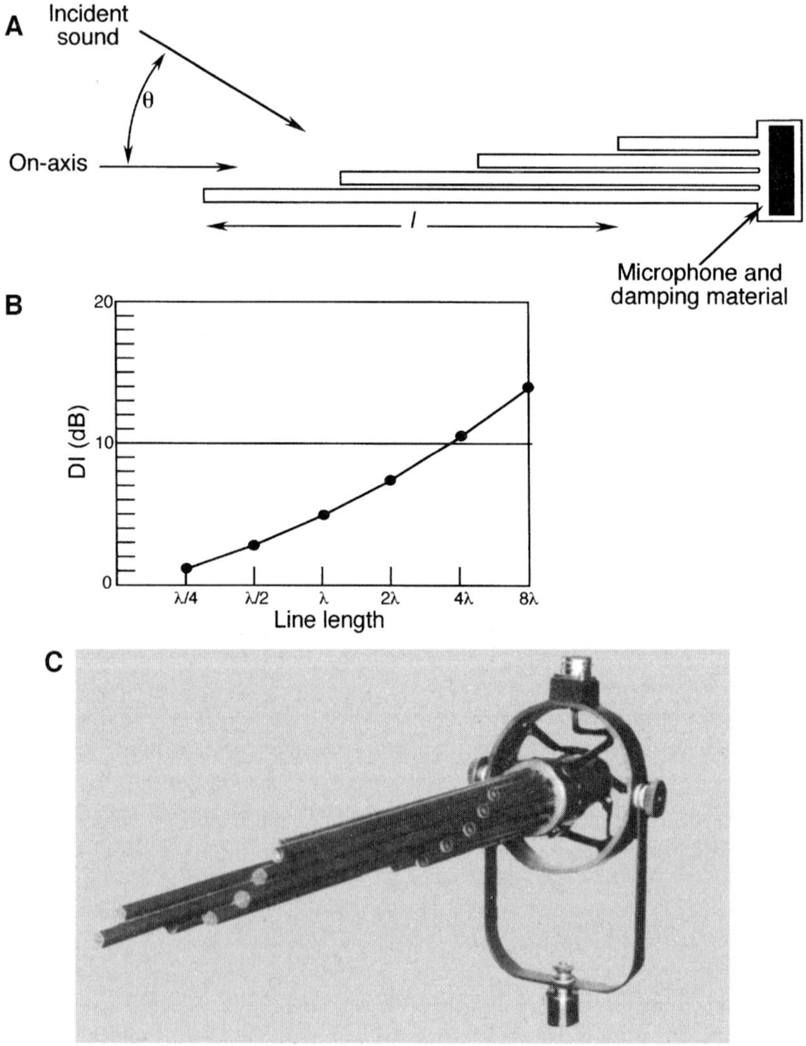

FIGURE 6–1

Olson's multiple tube directional microphone; physical view (A); plot of DI as a function of overall tube length and frequency (B); photo of a multiple tube directional microphone (C). (Data after Olson, 1957.)

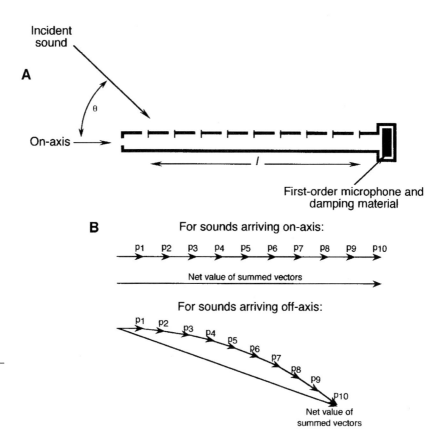

FIGURE 6-2

A single tube line
microphone (A); vector
diagram for on- and
off-axis operation (B).

The approximate directivity index (DI) for the array of tubes is shown in
Figure 6–1B as a function of array length and received signal wavelength.
For example, at 1 kHz the wavelength is approximately 0.3 m (12 in).
Thus, for an array length of 0.3 m the DI is about 5 dB. A photo of one
embodiment of the design is shown as C.

To a large extent, the complexity of the multi-tube array can be
replaced by the far simpler design shown in Figure 6–2A. Here, we can
imagine that the individual tubes have been superimposed on one another,
forming a single tube with multiple slot openings along its length. Such a
microphone as this is often termed a *line* microphone. In order to work,
the acoustical impedance at each opening must be adjusted so that sound,
once having entered the tube, does not readily exit at the next opening,
but rather propagates along the tube in both directions. Some designs
make use of external protruding baffles, or fins, which provide HF
resonant cavities that help to maintain off-axis HF response.

Because of its obvious physical appearance, the single-tube line microphone is commonly referred to as a "rifle" or "shotgun" microphone, and
it is in this form that most commercial high-directionality microphones are
designed today. As a general rule, the line section of the microphone is

added to a standard hypercardioid microphone, and it acts to increase the DI of the hypercardioid above a frequency inversely proportional to the length of the line.

A vector representation of the operation of the line microphone is shown in Figure 6–2B. For on-axis operation ($\theta = 0°$) the individual signal vectors are in phase and add to create a reference response vector, p. For sounds arriving slightly off-axis the individual vectors will be slightly displaced from each other in phase angle, and the vector summation will exhibit a reduction in level along with a lagging phase angle, relative to on-axis arrival. For sounds arriving considerably off-axis the angular displacement of individual phasors will be greater, and the vector summation will be even less.

These observations are borne out in measurements on commercial microphones. Figure 6–3A shows the AKG Model C 468 line microphone, which is available with one line section or with two, at the user's choice. The length of the microphone with a single section is 17.6 cm (7 in), and the length with two sections is 31.7 cm (12.5 in). Corresponding DI plots are shown at B. It can be seen that the DI data broadly follows that shown in Figure 6–4A.

The audio engineer in search of a high-directionality line microphone must perform the relevant calculations to make sure that a given microphone model is appropriate to the task at hand. The shorter models on the order of 30 cm (11 in) may produce the desired directionality at 4 kHz and above, but to maintain this degree of directionality at, say, 700 Hz, would require a line microphone some two meters in length! Not many line microphones have been designed to meet these requirements. One notable line microphone from the 1960s, the ElectroVoice Model 643 "Cardiline," had a length of 2.2 m (86 in). Many commercial line microphones have an operating length of little more than 0.2 m (8 in). Obviously, an operating point somewhere between these extremes will yield the best combination of directional performance and convenience of field operation and flexibility.

ESTIMATING THE PERFORMANCE OF A LINE MICROPHONE

The DI of a line microphone at low frequencies is simply the DI of the basic transducing mechanism, normally a hypercardioid element with its characteristic DI of 6 dB. As frequency increases, there is a point at which the line takes over and above which the DI tends to rise approximately 3 dB per doubling of frequency. That approximate frequency is given by Gerlach (1989):

$$f_0 = c/2l \qquad\qquad (6.2)$$

where c is the speed of sound and l is the length of the line section of the microphone (c and l expressed in the same system of units). Figure 6–4A presents a graph showing the rise in DI as a function of frequency for a line microphone of any length l. The relation between f_0 and l is given in

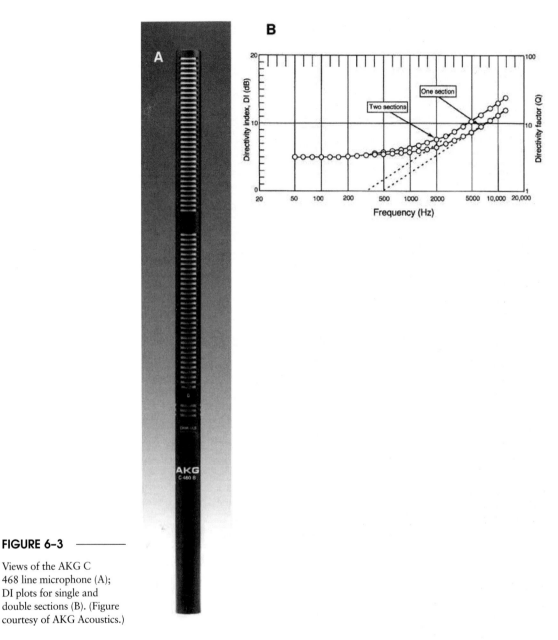

FIGURE 6-3

Views of the AKG C
468 line microphone (A);
DI plots for single and
double sections (B). (Figure
courtesy of AKG Acoustics.)

equation (6.1) and, for a given microphone, f_0 may be read directly from
the data shown in Figure 6–4B. For example, assume that a line microphone has a line section that is 200 mm (8 in) long. From Figure 6–4B we
can read directly the value of 900 Hz for f_0. Now, examining Figure
6–4A, we note that, if the microphone has a base hypercardioid element,
the effect of the line will only become apparent above about $3f_0$, or about
2700 Hz.

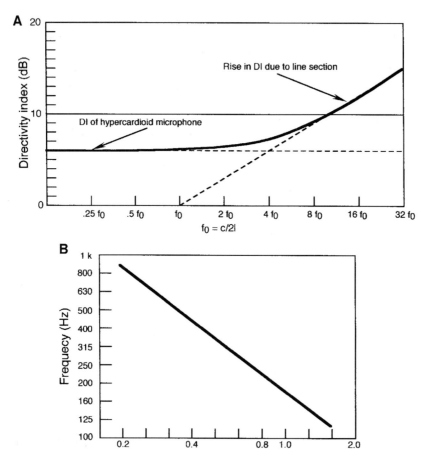

FIGURE 6-4

Directional performance
of a line microphone;
directivity index of a line
microphone as a function
of line length (A);
determining the value
of f_0 as a function of line
length, l (B).

When microphone directional performance is presented directly by
way of polar plots, the DI may be estimated by noting the included angle
between the $-6\,dB$ beamwidth values and entering that included angle
into the nomograph shown in Figure 6–5. (Note: *Beamwidth* is defined
here as the angular width of microphone pickup over which the response
loss is no greater than $-6\,dB$, relative to the on-axis value.)

PARABOLIC REFLECTORS AND ACOUSTIC LENSES

Figure 6–6A shows a section view of a parabolic reflector microphone,
as described by Olson (1957). Polar response is shown in Figure 6–6B,
and DI is shown at C. It can be appreciated how cumbersome such
microphones are to transport and use. However, microphones of this
type are used in field nature recording activities as well as in certain
surveillance activities, because the directional beamwidth at very high
frequencies can be quite narrow.

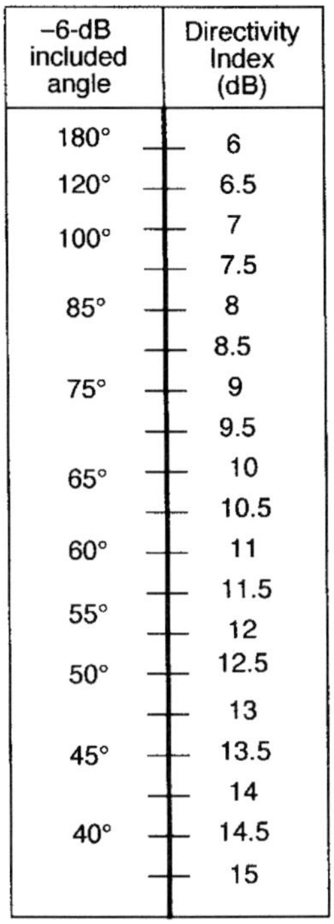

−6-dB included angle	Directivity Index (dB)
180°	6
120°	6.5
100°	7
	7.5
85°	8
	8.5
75°	9
	9.5
65°	10
	10.5
60°	11
	11.5
55°	12
50°	12.5
	13
45°	13.5
	14
40°	14.5
	15

FIGURE 6–5 ———

Determining approximate directivity index (DI) when the nominal −6 dB beamwidth of the microphone is given; for a given polar pattern, strike off the included angle over which the response is no less than −6 dB relative to 0 dB on-axis; then, using the nomograph, read directly the DI in dB.

A microphone based on the acoustic lens principle is shown in Figure 6–7 (Olson, 1957). Here, the lens acts as a converging element, focusing parallel rays of sound onto a transducer located at the focus of the lens. It is clear that a microphone such as this is of little practical use, and we show it primarily as an example of the acoustical lens principle.

SECOND- AND HIGHER-ORDER MICROPHONES

As we have seen in previous chapters, the basic family of directional microphones are referred to as first-order, inasmuch as the directional response is proportional to the first power of a cosine term. A second-order response pattern has directivity that is proportional to the square of the cosine term. Stated differently, a first-order microphone has response that is proportional to the pressure gradient, whereas a second-order design has response proportional to the gradient of the gradient.

The principle of the second-order microphone is developed as shown in Figure 6–8. At A, we show the physical circuit of a first-order cardioid microphone as a basis for comparison. The directional response of this microphone is given as $\rho = 0.5(1 + \cos \theta)$, where θ represents the bearing angle of sound arriving at the microphone.

If two such first-order microphones are placed very close together and their outputs subtracted from each other, we have their equivalent physical circuit as shown at B, and the directional response will be $\rho = (0.5 \cos \theta)(1 + \cos \theta)$. The directional pattern is shown at C in both linear and log (decibel) polar plots.

As we have seen in earlier chapters, the effective gradient distance between front and back openings in a first-order microphone *(D)* is quite small, perhaps no more than a centimeter or so. The requirement for an additional gradient distance, D', calls for extreme measures if this distance is to be minimized. Effective second-order microphones can be made to work out to perhaps 6 or 8 kHz, with diminished second-order effect above that point.

The general directional equation for a second-order microphone is:

$$\rho = (A + B \cos \theta)(A' + B' \cos \theta) \qquad (6.3)$$

where $A + B = 1$ and $A' + B' = 1$.

At present, second-order microphones are rarely used in the recording studio, and their major application is in the area of close-speaking, noise canceling operations in difficult communications environments. Proximity effect is a problem in that it rises at a rate of 12 dB/octave at low frequencies, rendering these microphones very sensitive to wind effects and close placement to sound sources. Woszczyk (1984) discusses some studio applications in detail.

As examples of second-order response, we show pertinent data for two designs: $\rho = (0.5 + 0.5 \cos \theta)(0.5 + 0.5 \cos \theta)$ (see Figure 6–9) and $\rho = \cos^2 \theta$ (see Figure 6–10).

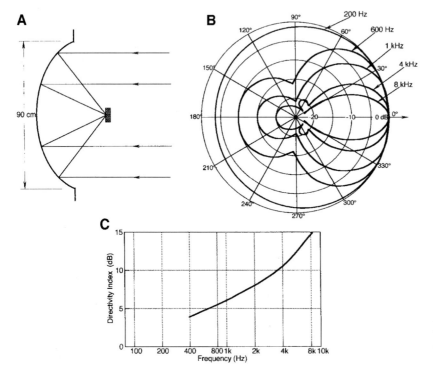

A

90 cm

B

C

FIGURE 6–6

Details of a parabolic
microphone; section
view (A); polar response (B);
directivity versus
frequency (C).

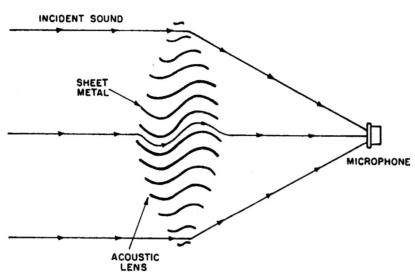

INCIDENT SOUND

SHEET
METAL

MICROPHONE

ACOUSTIC
LENS

FIGURE 6–7

Section view of an
acoustic lens directional
microphone.

The design shown in Figure 6–9 resembles a cardioid pattern with a
slightly narrowed contour; its response at ±90° is −12 dB, as compared
to −6 dB for a first-order cardioid. The physical circuit is shown at A,
and the actual realization is shown at B. Polar plots, both linear and log,
are shown at C.

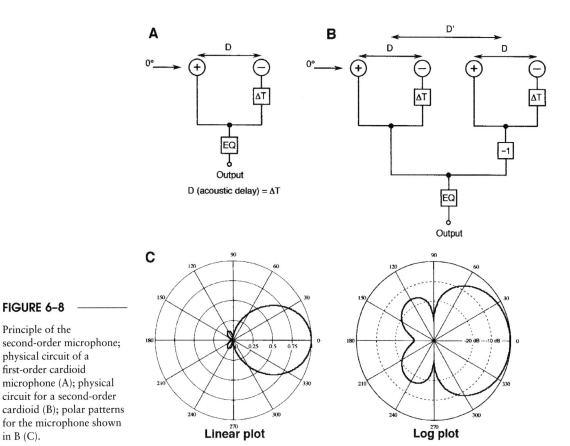

FIGURE 6-8

Principle of the second-order microphone; physical circuit of a first-order cardioid microphone (A); physical circuit for a second-order cardioid (B); polar patterns for the microphone shown in B (C).

The design shown in Figure 6–10 resembles a figure-8 pattern with a slightly narrowed contour, and its response at ±45° is 6 dB, as compared to −3 dB for the first-order figure-8 pattern. The physical circuit is shown at A, and the actual realization is shown at B. Polar plots, both linear and log, are shown at C.

VARIATIONS IN SECOND-ORDER DESIGN

The design of higher-order microphones can be simplified to some degree by sharing of elements, as shown in Figure 6–11. Here, the two sections making up the second-order design consist of elements one and two, and elements two and three. Element two is shared between both sections. In general, this approach provides a closer overall spacing between elements of the microphone. Here, note that distance D is the same for both gradient values.

Another design option is to limit the HF range of second-order action, and letting normal HF pattern narrowing to take over at higher frequencies, as shown in Figure 6–12. Here, a pair of cardioid elements are connected for second-order operation up to about 7 kHz. Above that

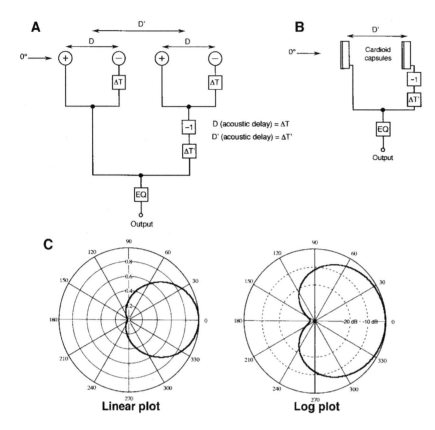

FIGURE 6-9

Design data for the second-order microphone, $\rho = (0.5 + 0.5 \cos \theta)(0.5 + 0.5\cos \theta)$: physical circuit (A); mechanical circuit (B); polar response (C).

frequency the contribution of the rear element is limited, and the front element's increasing directivity is allowed to take over response in the upper octave. The baffle surrounding the front element can be sized and shaped to optimize the transition from second-order response to beaming first-order response. Above about 12 kHz, the increased beaming of the front element can be ignored.

THE ZOOM MICROPHONE

The microphone shown in Figure 6–13 is intended to be used with a handheld video camera with a zoom lens. The microphone "zooms" electronically in synchronism with the lens system. The microphone array consists of three first-order cardioid elements whose outputs are combined in several ways to produce three distinct patterns, including the intervening patterns. The three primary patterns are:

1. Wide angle: omnidirectional pickup (sum of elements 2 and 3). It is produced when potentiometers R_1 and R_2 are in position 1.

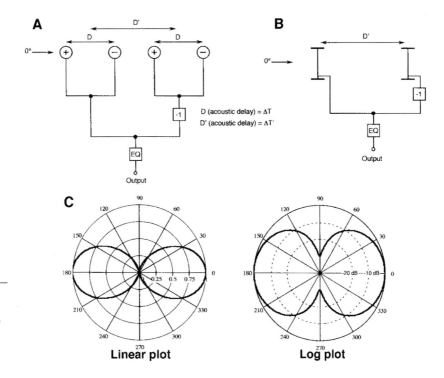

FIGURE 6-10

Design data for the
second-order microphone,
$\rho = \cos^2 \theta$: physical circuit
(A); mechanical circuit
(B); polar response (C).

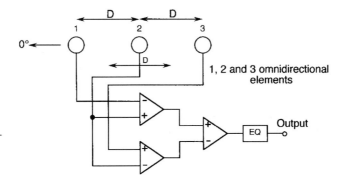

FIGURE 6-11

Second-order microphone
with shared elements.

2. Medium angle: cardioid pickup (element 1 alone). It is produced
   when potentiometers R_1 and R_2 are in position 2.

3. Narrow angle: second-order pickup (sum of elements 2 and 3 with
   necessary equalization). It is produced when potentiometers R_1 and
   R_2 are in position 3.

There is an overall gain shift of 12 dB throughout the microphone's
operating range to compensate for distance effects.

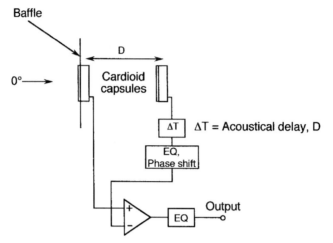

FIGURE 6-12

A combination of first- and
second-order microphone
performance.

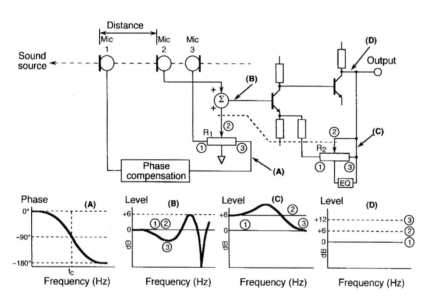

FIGURE 6-13

Details of a "zoom"
microphone. (Data after
Ishigaki et al., 1980.)

A THIRD-ORDER MICROPHONE

The microphone shown in Figure 6–14A has the polar equation
$\rho = \cos^3\theta$. The resulting directional response is shown at B. Third-order
microphones have been used primarily in noise canceling applications,
where their immunity to distant sound sources is considerable (Beavers
and Brown, 1970).

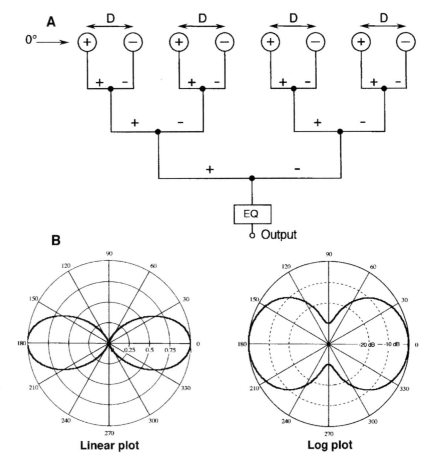

FIGURE 6-14

Details of a third-order microphone; physical circuit (A), polar response, both linear and log, (B).

CALCULATION OF DIRECTIVITY DATA FROM PATTERNS GIVEN IN POLAR COORDINATES

The RE (random efficiency) of a microphone whose polar pattern is symmetrical about its principal axis is given by:

$$\text{RE} = \frac{1}{2}\int_0^{\pi} \sin\theta [f(\theta)]^2 \, d\theta \tag{6.4}$$

where θ is the response angle in radians and $f(\theta)$ is the response value (ρ) at angle θ. If $f(\theta)$ can be described by cosine relationships (as most standard patterns can), equation (6.4) leads to a definite integral that can be easily solved.

TABLE 6-1   Microphone directivity data in tabular form (orders 0 through 4)

Pattern Equation	Random Efficiency (RE)	Directivity Factor (DF)	Directivity Index (DI)
$\rho = 1$ (omni)	1	1	0 dB
$\rho = \cos\theta$ (figure-8)	0.33	3	4.8
$\rho = \cos^2\theta$ (2nd order figure-8)	0.20	5	7.0
$\rho = \cos^3\theta$ (3rd order figure-8)	0.14	7	8.5
$\rho = \cos^4\theta$ (4th order figure-8)	0.11	9	9.5
$\rho = 0.5 + 0.5\cos\theta$ (cardioid)	0.33	3	4.8
$\rho = 0.25 + 0.75\cos\theta$ (hypercardioid)	0.25	4	6.0
$\rho = (0.5 + 0.5\cos\theta)\cos\theta$ (2nd order cardioid)	0.13	7.5	8.8
$\rho = (0.5 + 0.5\cos\theta)\cos^2\theta$ (3rd order cardioid)	0.086	11.6	10.6
$\rho = (0.5 + 0.5\cos\theta)\cos^3\theta$ (4th order cardioid)	0.064	15.7	12.0

Data from Olson, 1972

Data for several orders of microphone patterns are given in the Table 6–1.

The relationships among RE, DF, and DI are:

$$DF = 1/RE \qquad (6.5)$$
$$DI = 10 \log DF \text{ dB} \qquad (6.6)$$

C H A P T E R   7

MICROPHONE MEASUREMENTS, STANDARDS AND SPECIFICATIONS

INTRODUCTION

In this chapter we discuss the performance parameters of microphones that form the basis of specification documents and other microphone literature. While some microphone standards are applied globally, others are not, and this often makes it difficult to compare similar models from different manufacturers. Some of the differences are regional and reflect early design practice and usage. Specifically European manufacturers developed specifications based on modern recording and broadcast practice using capacitor microphones, whereas traditional American practice was based largely on standards developed in the early days of ribbon and dynamic microphones designed originally for the US broadcasting industry. Those readers who have a special interest in making microphone measurements are referred to the standards documents listed in the bibliography.

PRIMARY PERFORMANCE SPECIFICATIONS

1. Directional properties: Data may be given in polar form or as a set of on- and off-axis normalized frequency response measurements.

2. Frequency response measurements: Normally presented along the principal (0°) axis as well as along 90° and other reference axes.

3. Output sensitivity: Often stated at 1 kHz and measured in the free field. Close-talking and boundary layer microphones need additional qualification. Some manufacturers specify a load on the microphone's output.

4. Output source impedance.

5. Equivalent self-noise level.

6. Maximum operating sound pressure level for a stated percentage of total harmonic distortion (THD).

Additionally, a complete listing of mechanical and physical characteristics and any switchable performance features built into the microphone are described in this chapter.

FREQUENCY RESPONSE AND POLAR DATA

Frequency response data should always state the physical measuring distance so that an assessment of proximity effect in directional microphones can be correctly made. If no reference is made, it can be assumed that measurements are made at 1 meter. Data for professional microphones may be presented with tolerance limits, as shown in Figure 7–1. Here, the data indicate that the microphone's response falls within a range of ±2 dB above about 200 Hz (slightly greater below that frequency); however, there is no indication of the actual response of a sample microphone.

If the data can be presented with clarity, some manufacturers will show proximity effects at distances other than the reference one meter, as shown in Figure 7–2. This data is especially useful for vocal microphones that are intended for close-in applications.

Many manufacturers show response at two or more bearing angles so that the variation in response for those off-axis angles can be clearly seen, as shown in Figure 7–3. Here, the response for a cardioid is shown on-axis and at the nominal null response angle of 180°. For supercardioid and hypercardioid microphones, the response at the null angles of 110° and 135° may also be shown.

Taking advantage of normal microphone symmetry, polar plots may be restricted to hemispherical representation, as shown in Figure 7–4. For microphones that are end-addressed, it is clear that response will be

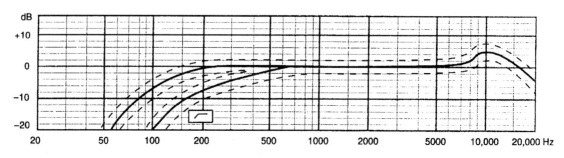

FIGURE 7–1

Amplitude response versus frequency with upper and lower limits for a capacitor vocal microphone; effect of LF cut is also shown. (Figure courtesy of Neumann/USA.)

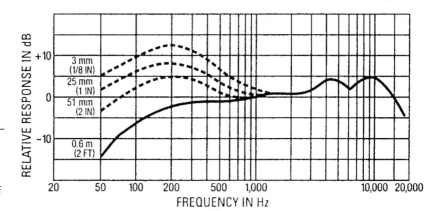

FIGURE 7-2

Proximity effect for a dynamic vocal microphone shown at several working distances. (Figure courtesy of Shure Inc.)

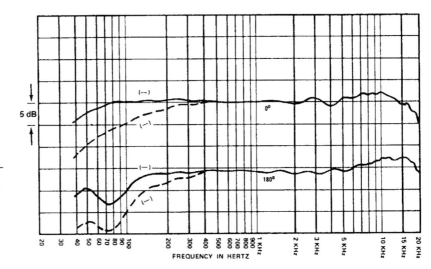

FIGURE 7-3

Amplitude response shown at reference angles of 0° and 180° for a Variable-D dynamic microphone. (Figure courtesy of Electro-Voice.)

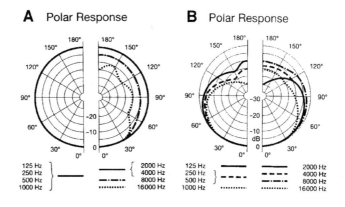

FIGURE 7-4

Microphone polar groups, hemispherical only, for omni capacitor (A) and cardioid capacitor microphones (B). (Figure courtesy of AKG Acoustics.)

symmetrical about the axis throughout the frequency band. However, for side-addressed microphones, there will be variations, especially at higher frequencies where the asymmetry in diaphragm boundary conditions becomes significant. Normally, this data is not shown, but there are interests within the microphone user community to standardize the presentation of more complete directional information. As present, such additional data is presented at the discretion of each manufacturer.

MICROPHONE PRESSURE SENSITIVITY

The principal method of presenting microphone sensitivity is to state the output rms voltage (mV/Pa) when the microphone is placed in a 1 kHz free progressive sound field at a pressure of 1 Pa rms (94 dB L_p). A nominal microphone load impedance of 1000 ohms may be stated as well, but the standard is normally referred to as the "open circuit" output voltage of the microphone. Another way of stating this data is to give the rms voltage output level in dB relative to one volt:

$$\text{Output level (dBV)} = 20 \log (\text{rating in mVrms}) - 60 \text{ dB} \quad (7.1)$$

MICROPHONE OUTPUT POWER SENSITIVITY

Microphone power output specifications were developed during the early days of broadcast transmission when the matched impedance concept was common. Here, the microphone is loaded with an impedance equal to its own internal impedance, as shown in Figure 7–5A. When unloaded, as shown at B, the output voltage is doubled.

The rating method is somewhat complicated, and we now give an example: consider a dynamic microphone with rated impedance of 50 ohms and an open-circuit output sensitivity of 2.5 mV/Pa. In modern specification sheets this voltage level may also be expressed as −52 dB dBV (re 1 volt). The same microphone, if its loaded output power is

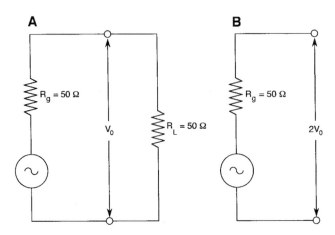

FIGURE 7–5

Microphone output unloaded (A); loaded (B).

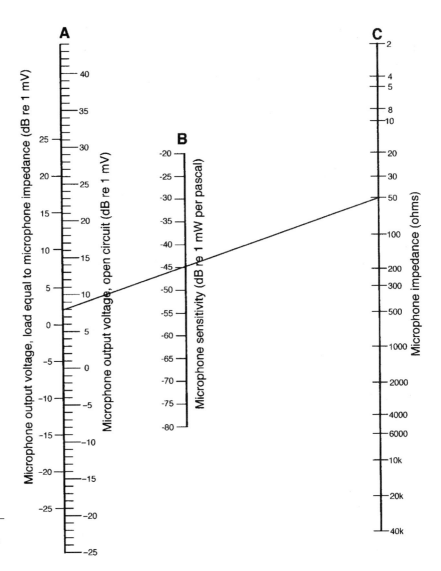

FIGURE 7-6

Microphone power output nomograph.

given, would carry an output power rating of −45 dBm. This is solved as follows.

When the microphone is given a matching load of 50 ohms, its output voltage will be reduced by one-half, or 1.25 mV. The power in the load will then be:

$$\text{Power} = (1.25)^2/50 = 3.125 \times 10^{-8} \text{ W, or } 3.125 \times 10^{-5} \text{ mW}$$

Solving for power level in dBm:

$$\text{Level} = 10 \log (3.125 \times 10^{-5}) = -45 \text{ dBm}$$

The nomograph in Figure 7–6 lets us solve this directly, as indicated by the line that has been drawn over the nomograph. Here, we simply take the unloaded output voltage level (re 1 mV) of +8 dB (60 − 52) and locate that value at A. The nominal impedance of the microphone (50 ohms) is located at B. A line is then drawn between the two points and the microphone's sensitivity, in dBm per pascal, is read directly at B.

Other rarely used variations of this method are:

1. Output in dBm per dyne per square centimeter (dBm measured in a matched load impedance at 74 dB L_P).
2. Output in dBm, EIA rating (dBm measured in a matched load impedance at an acoustical level of 0 dB L_P).

The reader can readily appreciate the simplicity and universality of the modern open circuit output voltage rating method.

MICROPHONE SOURCE IMPEDANCE

Virtually all of today's professional microphones, capacitor or dynamic, are what may be called "low impedance," as opposed to the high impedance models of decades past. The range of impedance may be typically from 50 to 200 ohms for capacitor microphones, or up to the range of 600 ohms for some dynamic models.

Since the traditional input load impedance seen by a microphone today is in the range of 2000 to 5000 ohms, it is clear that the load impedance is high enough that it has little measurable effect on the microphone's output voltage.

This being the case, the microphone's output impedance rating is of little practical consequence in modern systems layout. However, some microphone preamplifiers have a control for adjusting the input circuitry for specifically matching a wide range of microphone output impedances. (See Chapter 8 under the Stand-Alone Microphone Preamp.)

NORMAL RANGES OF MICROPHONE
DESIGN SENSITIVITY

In designing capacitor microphones the engineer is free to set the reference output sensitivity to match the intended use of the microphone. Table 7–1 gives normal sensitivity ranges.

The design criterion is simple; microphones intended for strong sound sources will need less output sensitivity to drive a downstream

TABLE 7–1   Normal sensitivity ranges by use

Microphone usage	Normal sensitivity range
Close-in, hand-held	2–8 mV/Pa
Normal studio use	7–20 mV/Pa
Distant pickup	10–50 mV/Pa

preamplifier to normal output levels, while distant pickup via boundary layer microphones or rifle microphones will need greater output sensitivity for the same purposes.

MICROPHONE EQUIVALENT SELF-NOISE LEVEL RATING

Today, the self-noise level of a capacitor microphone is expressed as an equivalent acoustical noise level stated in dB(A). For example, a given microphone may have a self noise rating of 13 dB(A). What this means is that the microphone has a noise floor equivalent to the signal that would be picked up by an ideal (noiseless) microphone if that microphone were placed in an acoustical sound field of 13 dB(A). Modern studio grade capacitor microphones generally have self-noise ratings in the range from 7 dB(A) to 14 or 15 dB(A). Tube models will have higher noise ratings, many in the range from 17 dB(A) to 23 dB(A).

As a practical matter, the self-noise of a modern capacitor microphone will be about 10 to 12 dB greater than the equivalent input noise (EIN) of a good console or preamplifier input stage; thus, the self-noise of the microphone will be dominant. With a dynamic microphone this is not normally the case; the output voltage of the dynamic microphone may be 10 to 12 dB lower than that of a capacitor model so that the EIN of the console will dominate. As a result of this, dynamic microphones do not carry a self-noise rating; rather, their performance must be assessed relative to the EIN of the following console preamplifier.

Some microphone specifications carry two self-noise ratings. One of these is the traditional A-weighted curve and the other is a psychometric curve that is more appropriate for microphones used in non-recording measurement purposes. The two curves are shown in Figure 7–7.

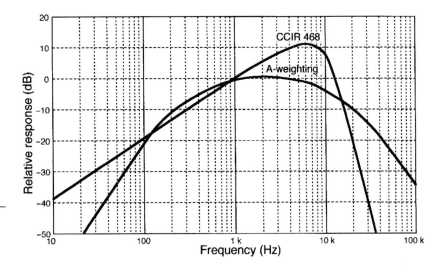

FIGURE 7–7

Two self-noise weighting curves for microphone measurements.

DISTORTION SPECIFICATIONS

For studio quality microphones the reference distortion limit is established as the acoustical signal level at 1 kHz which will produce no more than 0.5% THD (total harmonic distortion) at the microphone's output. Reference distortion amounts of 1% or 3% may be used in qualifying dynamic microphones for general vocal and hand-held applications.

Microphone distortion measurements are very difficult to make inasmuch as acoustical levels in the range of 130 to 140 are required. These levels are hard to generate without significant loudspeaker distortion. A pistonphone (mechanical actuator) arrangement can be used with pressure microphones where a good acoustical seal can be made, but it is useless with any kind of gradient microphone.

It has been suggested (Peus 1997) that microphone distortion measurements can be made using a twin-tone method in which two swept frequencies, separated by a fixed frequency interval, such as 1000 Hz, be applied to the microphone under test. Since the individual sweep tones are separately generated, they can be maintained at fairly low distortion; any difference tone generated by the diaphragm-preamplifier assembly represents distortion and can be easily measured with a fixed 1000 Hz filter, as shown in Figure 7–8. One problem with this method is that it is difficult to establish a direct equivalence with standard THD techniques.

In many studio quality microphones, the distortion present at very high levels results not from the nonlinearities of diaphragm motion but rather from electrical overload of the amplifier stage immediately following the diaphragm. Accordingly, some manufacturers simulate microphone distortion by injecting an equivalent electrical signal, equal to what the diaphragm motion would produce in a high sound field, and then measure the resulting electrical distortion at the microphone's output. This method assumes that the diaphragm assembly is not itself producing distortion, but rather that any measured distortion is purely the result of electrical overload. We must rely on the manufacturers themselves to ensure that this indeed the case.

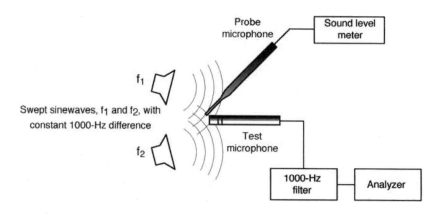

FIGURE 7–8

Twin-tone method of measuring microphone distortion.

MICROPHONE DYNAMIC RANGE

The useful dynamic range of a microphone is the interval in decibels between the 0.5% THD level and the A-weighted noise floor of the microphone. Many studio quality capacitor microphones have total dynamic ranges as high as 125 or 130, better by a significant amount than that of a 20-bit digital recording system.

As a matter of quick reference, many microphone specification sheets present nominal dynamic range ratings based on the difference between the A-weighted noise floor and a studio reference level of 94 dB L_P. Thus, a microphone with a noise floor of 10 dB(A) would carry a dynamic range rating of 84 dB by this rating method. No new information is contained in this rating, and its usefulness derives only from the fact that 94 dB L_P represents a normal maximum operating level in a broadcasting studio. Many manufacturers ignore this rating altogether.

MICROPHONE HUM PICKUP

While all microphones may be susceptible to strong "hum" fields produced by stray ac magnetic fields at 50 or 60 Hz and its harmonics, dynamic microphones are especially susceptible because of their construction in which a coil of wire is placed in a magnetic flux reinforcing iron yoke structure. Examining current microphone literature shows that there is no universally applied reference flux field for measuring microphone hum pickup. Several magnetic flux field reference standards may be found in current literature, including: 1 oersted, 10 oersteds, and 1 milligauss. The choice of units themselves indicates a degree of confusion between magnetic induction and magnetic intensity. The specification of hum pickup is rarely given today, perhaps due to the use of hum-bucking coils and better shielding in the design of dynamic microphones, as well as the general move away from tube electronics and associated power supplies with their high stray magnetic fields.

RECIPROCITY CALIBRATION OF PRESSURE MICROPHONES

Pressure microphones are primarily calibrated in the laboratory using the reciprocity principle. Here, use is made of the bilateral capability of the capacitor element to act as either a sending transducer or a receiving transducer, with equal efficiency in both directions. The general process is shown in Figure 7–9.

At A, an unknown microphone and a bilateral microphone are mounted in an assembly and jointly excited by a third transducer of unknown characteristics. From this measurement set we can obtain only the *ratio* of the sensitivities of the bilateral and unknown microphones.

The next step, shown at B, is to measure the output of the unknown microphone by driving it with the bilateral microphone, acting as a small loudspeaker. In this step we can determine the product of the two sensitivities, taking into account the electrical and acoustical equivalent circuits. The sensitivity of either microphone can then be determined by algebraic manipulation of the ratio and product of the sensitivities, taken on a frequency by frequency basis. As a secondary standard for microphone level calibration a *pistonphone* is normally used. The pistonphone is a mechanical actuator that can be tightly coupled to the capsule assembly of a pressure microphone and produces a tone of fixed frequency and pressure amplitude. Those interested in further details of microphone calibration are referred to Wong and Embleton (1995).

IMPULSE RESPONSE OF MICROPHONES

The impulse response of microphones is rarely shown in the literature because of the difficulties in achieving a consistent impulse source and interpreting the results. Usually, a spark gap discharge is used, but it has been shown that at high frequencies the spectrum is not consistent. Actually, the best results over the normal audio passband may be obtained using special loudspeaker mechanisms. Figure 7–10 shows the

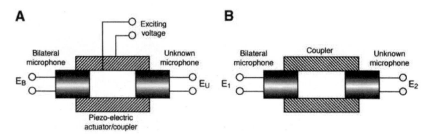

FIGURE 7-9

The reciprocity process; determining the ratio (A) and product (B) of microphone sensitivities.

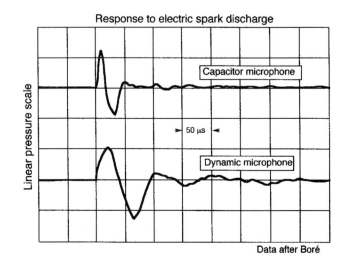

FIGURE 7-10

Impulse responses (spark gap) of capacitor and dynamic microphones. (Figure after Boré, 1989.)

spark gap response of both a capacitor and a dynamic microphone, and it can clearly be seen that the capacitor is better behaved in its time domain response.

Given future standardization of an adequate generating source, impulse response may become an important microphone specification, inasmuch as it presents more information regarding HF behavior of the microphone than is given by the traditional frequency response amplitude curve.

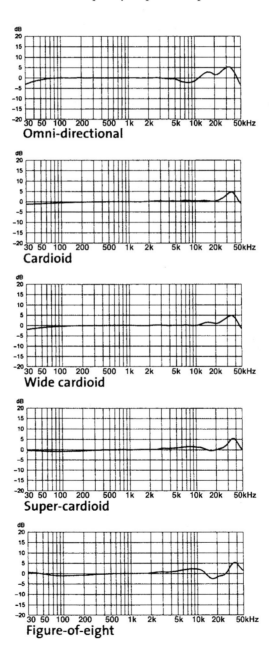

FIGURE 7-11

On-axis frequency response of Sennheiser model MKH800 microphone for five directional settings. (Figure courtesy of Sennheiser Electronics.)

EXTENDING THE MICROPHONE FREQUENCY
RESPONSE ENVELOPE

As higher sampling rates have been introduced into modern digital
recording, the need for microphones with extended on-axis frequency
response capability has increased. Figure 7–11 presents on-axis response
curves for the Sennheiser multipattern model MKH800 microphone
showing response out to 50 kHz – an example of what can be accom-
modated through careful design and amplitude equalization.

STANDARDS

The primary source of microphone standards for normal audio engineering
applications is the International Electrotechnical Commission (IEC). IEC
document 268-4 (1972) specifically lists the characteristics of microphones to be included in specification literature, along with methods for
performing the measurements. IEC document 327 (1971) specifically
covers reciprocity calibration of 1-inch pressure microphones.

In addition, many countries have their own standards generating
groups which in many cases adopt IEC standards, often issuing them
under their own publication numbers. See Gayford (1994, Chapter 10) for
additional discussion of microphone standards.

As the new century gets under way we are seeing microphone standards working groups focus attention on matters of time-domain performance along with greater resolution and detail in presenting off-axis
coverage. The attempt here is to define, far better than before, those subjective aspects that characterize microphone performance in real-world
environments and applications. We eagerly await the outcome of these
activities.

C H A P T E R   8

ELECTRICAL CONSIDERATIONS AND ELECTRONIC INTERFACE

INTRODUCTION

In this chapter we examine details of electronic performance of microphones and their interface with the following preamplifiers and consoles. Subjects to be covered include: remote powering, microphone output/preamp input circuitry, the stand-alone microphone preamp, microphone cable characteristics and interference, and overall system considerations. A final section deals with capacitor microphones operating on the radio frequency (RF) transmission principle.

POWERING

Most modern capacitor microphones operate on 48 V *phantom powering* (also known as "simplex" powering) provided by the preamplifier or console input module. The basic circuitry for phantom powering is shown in Figure 8–1. Here, a master 48 V dc supply provides positive voltage to pins 2 and 3 through a pair of 6800 ohm resistors, while a ground return path is provided through pin 1. The signal is carried on pins 2 and 3 and is unaffected by the presence of identical dc voltages on pins 2 and 3. Today, pin 2 is universally designated as the "hot" lead; that is, a positive-going acoustical signal at the microphone will produce a positive-going voltage at pin 2. Pin 1 provides both the dc ground return path for phantom powering as well as the shield for the signal pair.

The circuit normally used with a transformerless microphone for receiving power from the source is shown in Figure 8–2A. Here, a split resistor combination is tapped for positive dc voltage which, along with the ground return at pin 1, is delivered to the microphone's electronics and to the capsule for polarizing purposes.

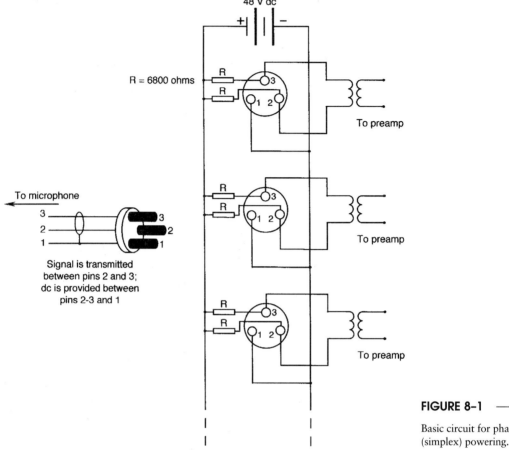

FIGURE 8-1 ——————

Basic circuit for phantom
(simplex) powering.

If the microphone has an integral output transformer, then a center-tapped secondary winding is used for receiving the positive voltage. This method is shown in Figure 8–2B. Phantom powering can also be used when there is neither a microphone output transformer or a console input transformer, as shown in Figure 8–2C.

Phantom powering at 48 V is normally designated as P48. There are also standards (IEC publication 268-15) for operation at nominal values of 24 V (P24) and 12 V (P12). Voltage tolerances and current limits for the three standards are shown in Table 8–1. The resistor values are generally held to tolerances of 1%.

T-powering (also known as A-B powering) is rarely encountered today. Circuit details are shown in Figure 8–3. It is normally designated as T12. T-powering is a holdover from earlier years and still may be encountered in motion picture work, where it is built into the many Nagra tape recorders used in that field. Here, the audio signal leads are at different dc voltages, and any residual hum or noise in the dc supply will be reflected through the microphone's output as noise.

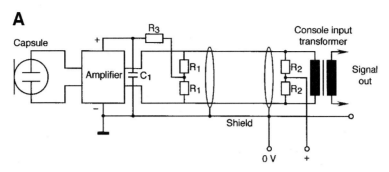

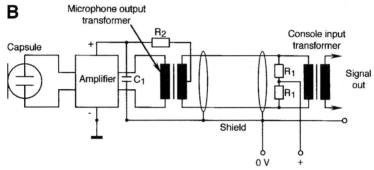

FIGURE 8-2

Microphone power input
circuitry using a resistive
voltage divider (A); using a
center-tapped transformer
secondary (B); using no
transformers in the
powering path (C).

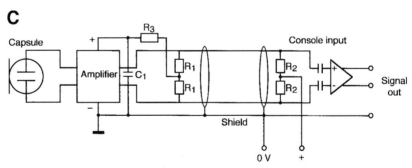

TABLE 8.1   Voltage tolerances and current limits

Supply voltage	12 ±1 V	24 ± 4 V	48 ± 4 V
Supply current	max. 15 mA	max. 10 mA	max 10 mA
Feed resistors	680 Ω	1200 Ω	6800 Ω

USING PHANTOM POWERING

Most microphone preamplifiers and console input sections have provision
for individual switching of phantom power on or off. When using dynamic
microphones it is good engineering practice to *turn off* the phantom powering, even though no current will flow through the microphone's voice
coil should the phantom power be left on. However, if T12 power is

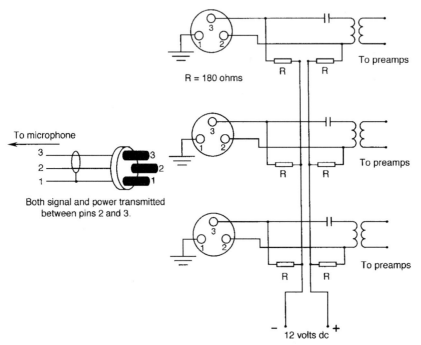

R = 180 ohms

To microphone

Both signal and power transmitted
between pins 2 and 3.

To preamps

To preamps

To preamps

12 volts dc

FIGURE 8-3

Basic circuit for
T-powering at 12 V.

inadvertently applied to a dynamic microphone, the 12 V dc will appear across the microphone's voice coil with noticeable deterioration of response and possible damage.

Another important rule is not to turn phantom power on or off when a microphone is bussed on and assigned to the monitor channels. The ensuing loud "pop" could easily burn out a HF transducer in the monitor loudspeaker systems. At the end of a session, it is normal practice to reduce both the master fader and the monitor level control to zero before shutting down all phantom power from the console.

While on the subject of phantom power, *never* attempt, when phantom power is on, to remove or replace the screw-on capsule that many capacitor microphones have. This has been known to burn out the FET in the input circuitry of the impedance converter.

Capacitor microphones vary in their susceptibility to shifts in nominal voltage in phantom powering. Generally, the variation is on the low side, such as may be encountered in very long microphone cable runs. Symptoms may be reduced signal output, increase in noise, as well as distortion. When these conditions occur with normal cable runs, the round-trip cable resistance should be measured, and the power supply itself checked for possible problems.

DC-TO-DC CONVERSION

Some capacitor microphones are designed to operate over multiple ranges of phantom powering, for example, from 20 to 52 volts dc to

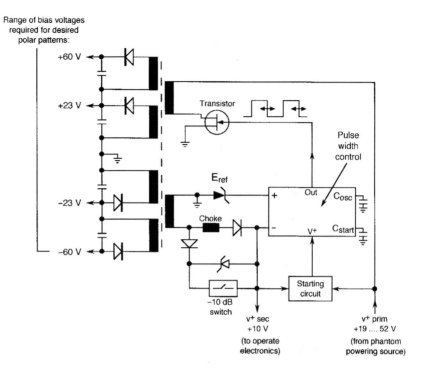

FIGURE 8-4

Details of dc-to-dc conversion for microphone operation at P24 and P48 standards. (Data after Neumann/USA.)

cover requirements of both P24 and P48 powering. What is required here is a circuit that converts the applied voltage to the value required for proper biasing of the capsule and operation of the impedance converting preamp. The circuit for the Neumann TLM107 microphone is shown in Figure 8–4. A major design challenge in such a circuit is the suppression of noise that could result from the high switching rate of the input voltage during the dc-to-dc conversion process.

This circuit provides capsule biasing voltages for its selectable patterns, reducing them accordingly when the $-10\,\text{dB}$ pad is engaged. The 10 V dc output is for powering the microphone's electronics.

RECENT DEVELOPMENTS IN PHANTOM POWERING

The present standard for phantom powering ensures that 48 V dc in a short circuit loading condition through two parallel 6800 ohm resistors will produce a current of 14 mA dc, thus limiting the current availability for a given microphone model. Some manufacturers have designed microphones that can accommodate greater current for handling higher sound pressure levels in the studio, and such microphones require lower resistance values in the phantom supply in order to receive the higher current. Generally, this has been carried out in proprietary stand-alone power supplies that a manufacturer may provide for a specific new microphone model.

Two-way compatibility is maintained. A new high-current microphone will work on a standard phantom supply, but it will not be able to attain its highest degree of performance. A standard microphone will work on the new supply, drawing only the current it needs.

Typical here is the "Super Phantom" powering that Josephson Engineering has specified for certain microphone models, in which current fed through a pair of 2200 ohm resistors in each leg of the power supply is directed to the microphone. A short circuit loading condition here would result in a current draw slightly in excess of 43 mA dc. International standardization activities are presently under way in this area.

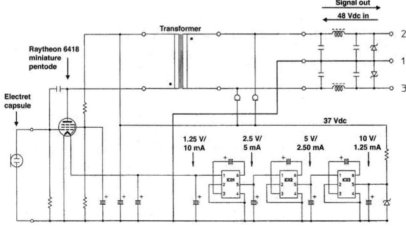

FIGURE 8-5

The Audio-Technica model AT3090 microphone; photo of microphone (A); circuit details (B). (Data courtesy of Audio-Technica USA.)

In an unusual design approach, Audio-Technica has introduced the model AT3060 tube-type microphone which is powered at P48. A photo of the microphone is shown in Figure 8–5A, and basic circuit details are shown at B. The Using a vacuum tube that will operate at a fairly low plate potential of 37 Vdc, the tube filament requirement of 1.25 Vdc at a current draw of 10 mA is attained via an unusual integrated circuit (IC) arrangement shown in the bottom-right portion of Figure 8–5B. The voltage at the input to the first IC is 10 V at a current of 1.25 mA. The three ICs in tandem progressively *halve* the voltage, while *doubling* the current, attaining a value of 1.25 V at a current draw of 10 mA at the output of the final IC. Various diodes are used for maintaining separation between signal and dc paths.

The microphone element itself is an electret. The AT3060 has a nominal sensitivity of 25.1 mV/Pa and can handle levels in the studio of 134 dB. The self-noise floor of the microphone is 17 dB(A).

Royer Labs has introduced integral electronics for phantom powering in two of their ribbon microphone models. Ribbons are particularly susceptible to variations in downstream loading, and, with their relatively low sensitivity, they may be subject to electrical interference in long runs. Figure 8–6A shows a photo of the Royer Labs R-122, with performance curves shown at B and C. The circuit diagram is shown in Figure 8–7.

The most unusual aspect of the circuit is the compound transformer, which has four parallel sets of windings resulting in an optimum impedance match to the ribbon. The secondaries of these four sections are connected in series to attain a voltage gain of 16 dB, looking into the very high input impedance of the buffer stage. The system has a maximum level capability of 135 dB L_P and a self noise no greater than 20 dB(A). The sensitivity is 11 mV/Pa.

POWER SUPPLIES FOR OLDER TUBE MICROPHONES

Classic tube-type capacitor microphones are as popular today in studio recording as they have ever been. In a typical power supply (one for each microphone), dc voltages are produced for heating the vacuum tube's filament, biasing the capsule, and providing plate voltage for the vacuum tube amplifier. In many dual diaphragm designs, remote pattern switching is also included in the supply. Such a design is shown in Figure 8–8. Note that the cable connecting the power supply to the microphone contains seven conductors.

MICROPHONE OUTPUT/CONSOLE PREAMP INPUT CIRCUITRY

THE MICROPHONE OUTPUT PAD

Most capacitor microphones have an integral output pad, as described in detail in Chapter 3, under Details of preamplifier and polarizing circuitry.

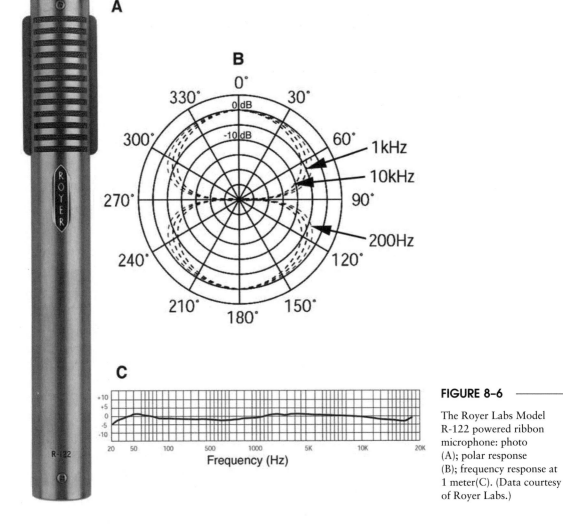

FIGURE 8-6

The Royer Labs Model R-122 powered ribbon microphone: photo (A); polar response (B); frequency response at 1 meter(C). (Data courtesy of Royer Labs.)

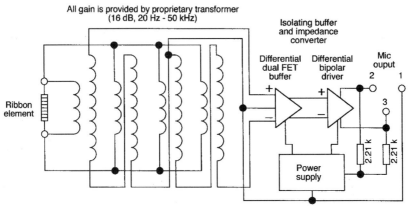

FIGURE 8-7

Circuit diagram for the Royer Labs Model R-122 microphone. (Figure courtesy of Royer Labs.)

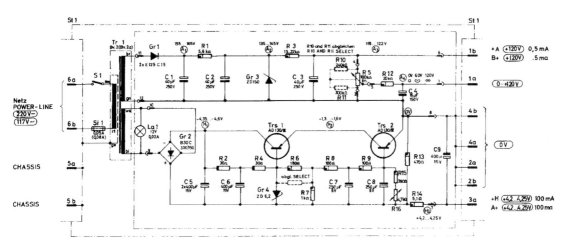

FIGURE 8-8

Power supply schematic for a tube capacitor microphone with variable pattern switching at the power supply; pin 1A provides the pattern control voltage. (Figure courtesy of Neumann/USA).

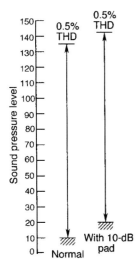

FIGURE 8-9

Typical effect of −10 dB microphone output pad on maximum level capability and self-noise of a capacitor microphone.

The effect of the pad is basically to shift the microphone's entire operating range from noise floor to overload point downward by some fixed amount, normally 10 to 12 dB. The effect of this is shown in Figure 8–9. Note that the total dynamic range of the microphone remains fixed, with or without the pad.

THE MICROPHONE OUTPUT TRANSFORMER

A number of capacitor microphone models have an integral output transformer built into the microphone body. Most early tube-type capacitor microphones had output transformers that were usually housed in the power supply unit. The transformer often had split secondary windings and could be strapped for either parallel or series operation, as shown in Figure 8–10. Typically, when the windings are strapped for parallel operation (A), the output impedance is 50 ohms; for series strapping (B) the output impedance is 200 ohms. In either case the output power capability remains the same:

$$\text{Output power} = E^2/R = V^2/50 = (2V)^2/200.$$

In many cases it is necessary to reduce the output signal from the microphone even lower than the parallel transformer strapping. H-pads (shown in Figure 8–10 C) may be used both to maintain the impedance relationship and produce the desired amount of attenuation. Today, most transistorized capacitor microphones are transformerless and operate at a fixed balanced output impedance.

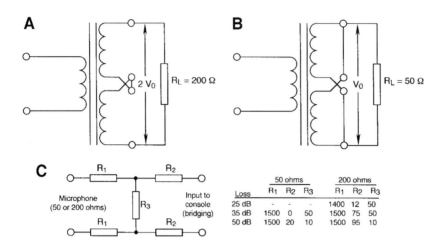

FIGURE 8–10

Transformer strapping for 200 ohms (A) and 50 ohms output impedance (B); H-pad values for balanced attenuation for microphone impedances of 50 and 200 ohms (C).

CONSOLE INPUT SECTION

In the very early days of broadcasting and recording, the dynamic microphones of the period had relatively low outputs and normally operated into matching input impedances. Typically, a microphone with a 600-ohm source impedance looked into a 600-ohm load, following the matched impedance concept. Impedance matching was a holdover from early telephone practice and found its way into broadcasting, and from there into recording. Today, the *bridging* concept is well established. In a bridging system, all output impedances are relatively low and all input impedances are relatively high throughout the audio chain. Here, the ratio of high to low impedance is normally in the range of 10-to-1 or greater.

Figure 8–11 shows a simplified transformerless console input section showing switchable line-microphone operation. At one time transformers were felt to be indispensable in the design of microphone input circuitry. Their chief advantages are a high degree of electrical balance and consequent high common-mode signal rejection. (A common-mode signal is one that is identical at both inputs; typically, induced noise signals are common mode.) In the era of vacuum tubes the input transformer was of course essential. Today's best solid state balanced input circuitry does not mandate the use of transformers, and there are considerable economic advantages to pass on to the user. Only under conditions of high electrical interference might their use be required.

Today, most microphones have an output impedance in the range of 50 to 200 ohms. Most consoles have an input bridging impedance in the range of 1500 to 3000 ohms.

IMPROPER MICROPHONE LOADING

A typical 250-ohm ribbon microphone may have an impedance modulus as shown in Figure 8–12A. As long as the microphone looks into a high

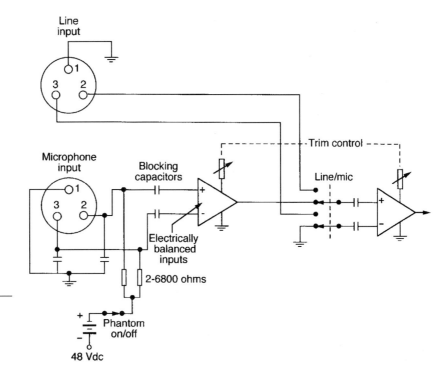

FIGURE 8-11

Simplified circuit for a transformerless console microphone/line input section.

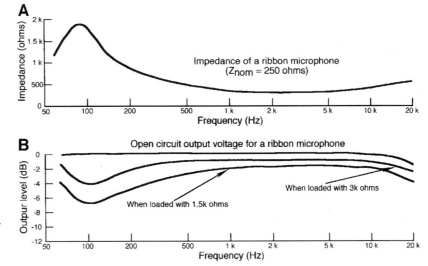

FIGURE 8-12

Effect of loading on ribbon microphone response.

impedance load, its response will be fairly flat. When used with a modern console having 1500- or 3000-ohm input impedances, the frequency response will be altered as shown at B. This is a response problem routinely encountered today and frequently goes unchallenged.

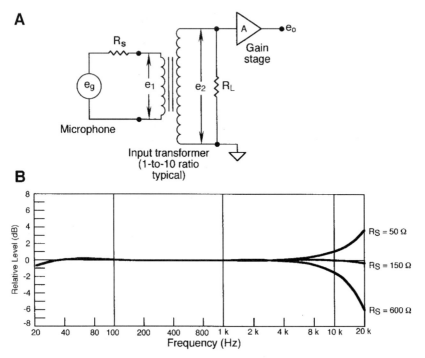

FIGURE 8-13

Response of transformer input circuit to varying values of microphone source impedance.

Another problem with improper loading is shown in Figure 8–13. Here, a capacitor microphone is loaded by a console input transformer that has a non-optimal termination on its secondary side. This can produce response variations similar to those shown in the figure as a function of the microphone's output impedance. Note that as the input impedance is reduced, the response develops a rise in the 10 kHz range. This comes as a result of an undamped resonance involving the stray capacitance between the primary and secondary windings of the transformer (Perkins, 1994).

UNBALANCED MICROPHONE INPUTS

Only in the lowest cost paging systems is one likely to come across an unbalanced microphone input. For fairly short cable runs from microphone to amplifier, there may be no problems. For longer runs, where there is greater likelihood for interference, the difference between balanced and unbalanced operation is as shown in Figure 8–14. For balanced operation, shown at A, induced noise will be equal and in-phase in both signal leads; it will be effectively canceled by the high common mode rejection of the input circuitry. For unbalanced operation, shown at B, the induced signal currents will be different between shield and conductor, and the noise will be significant.

MICROPHONE SPLITTERS

For many field operations, stage microphones must be fed to both recording and sound reinforcement activities. Microphone splitters are used to provide both an electrically direct feed to one operation, normally the recording activity, and a one-to-one transformer feed to the other activity. Circuit details for a typical passive splitter are shown in Figure 8–15. Here, there are two secondary windings, providing an additional output for broadcast activities. Ground lifts are often used to avoid hum due to

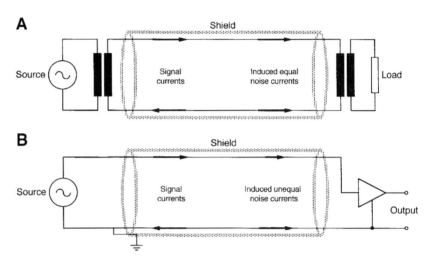

FIGURE 8–14

Balanced (A) versus unbalanced (B) microphone operation.

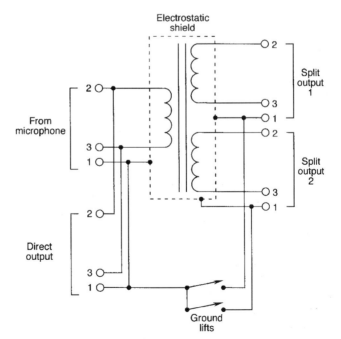

FIGURE 8–15

Details of a passive microphone splitter.

ground loops. When using passive splitters, it is essential that all loads fed by the splitter be bridging. Active splitters are often used in order to avoid expensive transformers and to permit feeding the signal to low input impedance loads.

THE STAND-ALONE MICROPHONE PREAMP

Many leading recording engineers prefer to bypass console microphone inputs altogether and use multiple individual stand-alone microphone preamps instead. These microphone preamps are normally highly refined versions of what may be found in the typical console, some offering significant improvements in areas of noise floor, common mode rejection, increased output level capability, source impedance matching, calibrated step-type gain trim, and rugged construction. Other features in some models include equalization options and metering. These performance attributes do not come cheaply, and a set of 16 individual preamps may cost many times more than an excellent mass-produced 24-input console.

For many recording activities it may be difficult to justify the expense of stand-alone preamps, considering that their outputs will be fed into a host console for normal signal routing and bus assignment. However, there are specific applications where their use is appropriate:

- Direct-to-stereo via two microphones, where no console routing functions are needed.
- Pop multichannel recording, with each microphone assigned to one recording channel. Here, the local host console is relegated to control

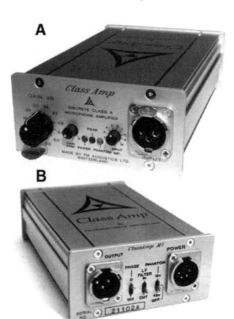

FIGURE 8-16

Front (A) and back (B) views of a stand-alone microphone preamplifier. (Figure courtesy of FM Acoustics.)

room and studio monitoring only, with all studio microphones and direct feeds going to their own preamps and from there to their individual recorder inputs. In this regard, the set of external stand-alone preamps has taken the place of the "channel path" in a modern in-line console.

Many producers and engineers prefer to work in this manner when called upon to go into an unknown studio to work with an artist who may be on tour and wishes to lay down tracks, or perhaps do additional work on an album in progress. One great benefit for the producer/ engineer is the comfort of working with a consistent set of tools that can be taken from one field location to another, reserving for a later time all post-production activities that will take place in a known, home environment.

Figure 8–16 shows front and back panel views of a high-quality stand-alone microphone preamplifier.

LINE LOSSES AND ELECTRICAL INTERFERENCE

To a very great extent, the recording engineer working in a well maintained studio does not have to worry about microphone cables, except to make sure that they are in good repair. Things may not be so simple for the engineer working in a specific remote location for the first time and finding that he may actually run short of cable! There are two concerns with long cable runs: phantom power operation and HF losses due to cable capacitance.

Typical electrical values for professional quality microphone cable are:

Gauge: #24 AWG stranded copper wire

Resistance/meter: 0.08 ohms

Capacitance/meter: 100 pF

P48 phantom powering has a fairly generous design margin for long cable runs. For example, a 100 m run of cable will have a resistance, per leg, of 8 ohms, and the total resistance in the cable will be 16 ohms. Considering the relatively high resistance of the phantom feed network of 6800 ohms per leg, this added value is negligible.

More likely, HF rolloff will be encountered in long cable runs, as shown in Figure 8–17. Here, data is shown for cable runs of 10 and 60 meters, with source impedances of 200 and 600 ohms.

Another problem with long cable runs is their increased susceptibility to RF (radio frequency) and other forms of electromagnetic interference. Local interference may arise from lighting control systems, which can generate sharp "spikes" in the power distribution system; these can be radiated and induced into the microphone cables. Likewise, nearby radio transmitters, including mobile units, can induce signals into microphone cables.

Cables of the so-called "starquad" configuration (developed early in telephony) can reduce interference by up to 10 dB relative to the normal two-conductor configuration. The starquad configuration is shown in Figure 8–18. Here, four conductors within the braided shield are twisted throughout the run of the cable. Diagonally opposite pairs are coupled and connected at each end of the cable to pins 2 and 3. The twisting of the pairs ensures that induced noise components are equal in each leg of the balanced signal pair, resulting in cancellation of noise components at the receiving end.

PHYSICAL CHARACTERISTICS OF CABLES

No engineer should ever stint on cable quality. The best cables available are of the starquad configuration, supple, and are easily coiled. In normal use, cables may be stepped on, crimped by doors and wheels on roll-about equipment, and otherwise subjected to daily abuse. In general, braided shield is preferable to wound foil shield; however, in permanent installations this may not be important.

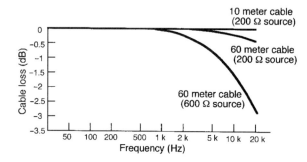

FIGURE 8–17

Effects of cable length and microphone impedance on HF response.

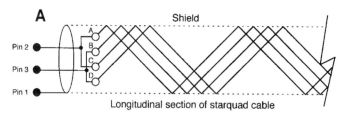

Longitudinal section of starquad cable

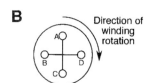

End view of cable showing internal rotationpattern of wire pairs

FIGURE 8–18

Details of starquad microphone cable construction.

Interconnecting hardware should be chosen for good fit, with an awareness that not all brands of connectors easily work together, even though they nominally meet the standards of XLR male and female receptacles.

Microphone "snakes" are made up of a number of individual microphone cables enclosed in one outer sheath, normally numbering 12 to 16 pairs. In an effort to keep the size down, foil inner shields are normally used on each pair. At the sending end it is preferable to have the snake terminate in a metal box with XLR female receptacles, rather than a fan-out of female XLR receptacles. This recommendation is based on ease of reassignment of microphones by cable number, should that be necessary. At the receiving end a generous fan-out should be provided for easy access to the console's inputs, with each cable number clearly and permanently indicated.

Capacitive coupling between pairs is generally quite low, and for signals at microphone level we can usually ignore it. However, for some applications in sound reinforcement, there may be microphone signals sent in one direction along the snake with concurrent line level monitoring signals sent in the opposite direction. This is not recommended in general recording practice.

CABLE TESTING

A microphone emulator circuit is shown in Figure 8–19. The circuit shown is single-ended (unbalanced) and should not necessarily be used to assess details of interference. However, it is excellent for determining HF cable losses over long runs. The 40-dB input pad will reduce an applied signal of, say 0.2 volts to 20 mV, a typical capacitor microphone sensitivity rating at 1 pascal. The 20 mV signal appears across a 200-ohm resistor and thus simulates the output of an actual microphone.

The circuit shown in Figure 8–20A is useful for simple cable continuity and leakage testing. Measurement across like pin numbers should result in a virtual short-circuit. Measurements across different pin numbers should be made with the ohmmeter set to a high resistance range in order to identify any stray leakage between wire pairs. The more sophisticated cable tester shown at *B* is typical of items offered by a number of companies. As an active device it can be used to test the following:

1. Cable continuity
2. Cable intermittent failure
3. Phantom power integrity
4. In-circuit continuity (via test tones)
5. Cable grounding integrity

Three types of cables can be tested using this system.

Polarity checking of a microphone/cable combination can be made using a system such as is shown in Figure 8–21. Here, a LF positive-going

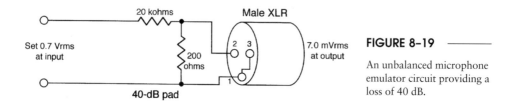

FIGURE 8-19 ────────

An unbalanced microphone
emulator circuit providing a
loss of 40 dB.

A   **B**

FIGURE 8-20 ───

Details of a simple continuity and leakage detector for microphone cables (A); a more complex cable testing apparatus (B).
(Photo B courtesy of Behringer Audio Technology.)

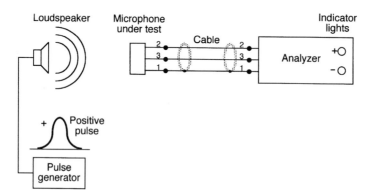

FIGURE 8-21 ────────

Details of a system for
checking signal polarity of
microphones.

acoustical pulse is fed to the microphone and analyzed at the other end
of the cable run using an analyzer that detects the presence of a positive-
or negative-going pulse. If negative polarity is indicated, then there is
near certainty that a cable has been miswired or that an item of in-line
equipment is inverting the signal. Virtually all modern amplifiers and signal processing devices are non-inverting. Thus, most electrical polarity
problems encountered today are the result of local wiring mistakes.

STAGE PREAMPLIFIERS AND FIBER OPTIC LINES

While normal phantom powering can easily handle cable runs up to the 100 meter range, there may be some environments where RF interference is a chronic problem. One option is to use stage microphone-to-line preamplifiers, which will raise the level of the transmitted signals by 40 or 50 dB with very low output impedance, thus providing considerable immunity to induced noise and losses due to long cable runs. Stage preamplifiers are available with remote controlled gain adjustment.

Another option is to provide on-stage digital conversion for each microphone and transmit the individual digital signals via fiber optic cables to a digital console in the control room. In this form, the signals may be sent over extremely long runs with no loss.

SYSTEM CONSIDERATIONS

GROUND LOOPS

One of the most common audio transmission problems is the *ground loop*. Figure 8–22 shows how a ground loop is created. Electronic devices are cascaded as shown. Note that there is a ground path, not only in the cables that connect the devices but also in the metal rack that houses them. Any ac power flowing in the rack will generate an external magnetic field, and that magnetic flux which flows through the loop will induce a small current through it.

It is necessary to break the continuity in the loop, and this is normally done as shown in Figure 8–23. Microphone cables are grounded at the

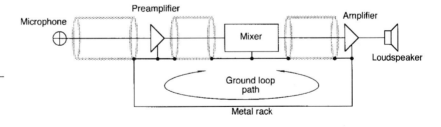

FIGURE 8-22

Generation of a ground loop.

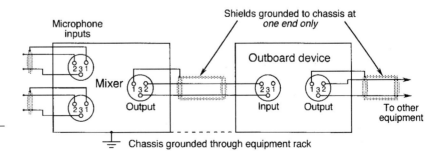

FIGURE 8-23

Breaking the ground loop.

input to the mixer in order to isolate the microphone from electrostatic (RF) interference and also to provide a continuity path for phantom powering. From the mixer onward, it is customary to connect the wiring shield at the output end only as shown. In this manner the continuity of the ground loop conductive path is broken, thus minimizing the severity of the ground loop.

GAIN STRUCTURE

Many audio transmission problems begin with the microphone and the initial input gain setting looking into the console. Modern consoles are well engineered, and essentially all the operator has to do to adjust the input and output faders to their nominal zero calibration positions and then set normal operating levels on each input channel using the input trim control. This may sound simple but it requires a little practice.

If the operator sets the microphone trim too low, then there is the risk that input noise in the console will become audible, perhaps even swamping out the noise of the microphone; if the trim is set too high, then there is the risk that the console's input stage will be overloaded on loud input signals. Since the console has a wider overall operating dynamic range than a microphone, the requirement is only to adjust the trim so that the program delivered from the studio via the microphone will fit comfortably within the total dynamic range of the console.

Figure 8–24 shows a level diagram for a single input channel through a typical console. Let us assume that we have a microphone whose sensitivity is 21 mV for an acoustical signal of 94 dB L_p and whose self noise floor is 10 dB(A). Further assume that the microphone's maximum output level (0.5% THD) is 135 adB L_p.

The microphone's output is equivalent to −31 dBu, where dBu is defined as:

$$dBu = 20 \log(\text{signal voltage}/0.775) \qquad (8.1)$$

A level of 94 dB L_p is typical of many instruments in the pop/rock studio when picked up at fairly close quarters and at normal playing levels, and thus the engineer will set the input trim control so that this electrical level will produce console output meter deflections in the normal operating range.

Note also that with this input setting the microphone's noise floor of 10 dB(A) will correspond to an electrical level 84 dB *lower* than −31 dBu, or −115 dBu. This level is about 13 dB below the noise floor of the console, which is in the range of −128 dBu. Thus, it is clear that the audio channel's noise floor will be essentially that of the microphone, with little contribution from the console's input circuitry. As the signal progresses through the remainder of the console, the noise floor does not change relative to normal operating level.

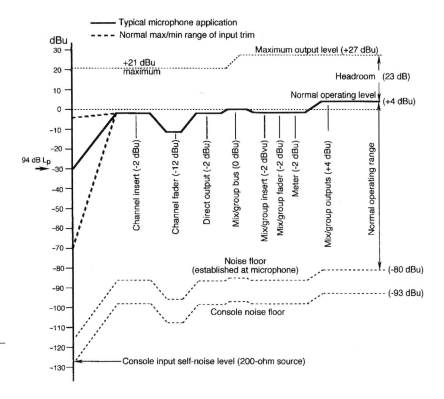

FIGURE 8-24

Typical console gain structure or level diagram.

The microphone input channel has the capability, for undistorted console output, of handling an input signal that is 23 dB higher than 94 dB, or 117 dB L_p. If the microphone's input signal exceeds this amount on a continuing basis, then the engineer must adjust the input trim to accommodate it. If the signal exceeds this amount only occasionally then the engineer may make necessary adjustments at the input fader.

The microphone's undistorted output level extends up to 135 dB L_P, which is 18 dB higher than the console's maximum output capability. As such, this microphone may be used in a wide variety of acoustical environments, and the only adjustments that need to be made are the microphone's output pad (for very loud signal conditions) or resetting the input trim as required.

SUMMING MULTIPLE INPUTS INTO A SINGLE OUTPUT CHANNEL

During remix operations, a number of microphones are often fed to a given output bus. For each new microphone input added to a given bus it is apparent that the overall input levels must be adjusted so that the signal fed downstream is uniform in level. Figure 8-25 shows the way that individual inputs should be adjusted so that their sum will be consistent, whether one or more microphone inputs are used.

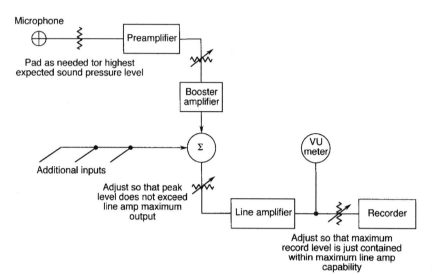

FIGURE 8-25

Combining multiple inputs.

Recording engineers engaged in direct-to-stereo recording, or in making stereo mixes from multitrack tapes, must be aware of the process discussed here.

MICROPHONES USING THE RF (RADIO FREQUENCY) TRANSMISSION PRINCIPLE

The capacitor microphone discussed in Chapter 3 operates on the principle of variable capacitance with fixed charge producing a variable signal output voltage. There is another method of deriving a signal from variable capacitance that has been successfully used by only a few microphone manufacturers, and that is the RF transmission principle. Today, only the Sennheiser company of Germany manufacturers such microphones on a large scale.

The RF capacitor microphone has exactly the same acoustical characteristics as a dc polarized microphone of the same physical geometry. The only difference is the method of converting the capacitance variations into a signal output.

Hibbing (1994) gives a complete account of the history and analysis of RF microphones. He describes two general methods, and simplified circuits for each are shown in Figure 8–26. The circuit shown at A works on the phase modulation (PM) principle and as such resembles a small FM transmitter-receiver combination in a single microphone package. The variable capacitance alters the tuning of a resonant circuit; adjacent to the tuned circuit are a discriminator section and a fixed oscillator operating in the range of 8 MHz. All three sections are mutually coupled through RF transformers. The alternations of tuning resulting from audio pressure variations affect the balance of upper and lower sections of the discriminator, and an audio signal output is present at the output.

Circuit details of the AM bridge design are shown in Figure 8–26B. Here, a dual-backplate push-pull diaphragm is used to advantage as a push-pull voltage divider, distributing the RF signal equally to both sides of the bridge circuit when in its rest position. Under audio excitation, the bridge output will be proportional to diaphragm excursion, providing a high degree of linearity. It is this modulation principle that is used in the Sennheiser MKH-20 series of studio microphones.

Early RF microphones were prone to internal instability and on occasion interference from local radio transmissions. Over the years the art has literally "fine-tuned" itself, and the modern Sennheiser MKH-series microphones are superb performers in all respects.

PARALLEL OPERATION OF MICROPHONES

While not generally recommended, it is possible to operate two microphones at a single console input if phantom power is not required. The price paid for this is a reduction in level of about 6 dB for each microphone and of course the inability to adjust levels individually. The circuit shown in Figure 8–27 shows how this can be done.

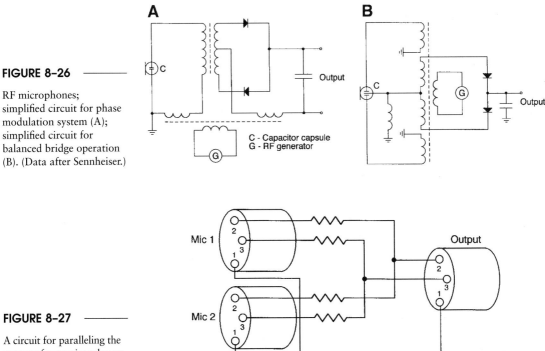

FIGURE 8-26

RF microphones; simplified circuit for phase modulation system (A); simplified circuit for balanced bridge operation (B). (Data after Sennheiser.)

FIGURE 8-27

A circuit for paralleling the output of two microphones. (Data after Shure Inc.)

DIGITAL MICROPHONES

So-called digital microphones have entered the marketplace during the last six or so years. Strictly speaking, these models are not actually "digital" in the specific sense of directly generating a digital output code from the diaphragm. Rather, they make use of traditional dc bias and preamplification of the analog signal at the diaphragm. It is only after this stage that analog-to-digital conversion takes place.

The advantage of these microphones is that certain problems in digital processing can be dealt with earlier, rather than later, in the audio chain. For example, the useful signal-to-noise ratio of a well-designed 25 mm (1 in) condenser diaphragm can be in the range of about 125 to 135 dB. An ideal 20-bit system is capable of a signal-to-noise range of 120 dB, and in a traditional recording system this will require truncation of the available dynamic range of the microphone by about 10 dB. In and of itself, this may or may not be a problem, depending on other electrical and acoustical considerations in the actual studio environment.

A

In the beyerdynamic model MCD100 series, the capsule looks into a 22-bit conversion system directly when the acoustical level is high (greater than 124 dB L_P). For normal studio levels (less than about 100 dB L_P), –10 or –20 dB padding can be inserted ahead of the digital conversion stage in order to optimize the bit depth. Sophisticated level control prevents the system from going into digital clipping. The microphone and associated signal flow diagram is shown in Figure 8–28A and B.

The Neumann Solution-D uses two 24-bit A-to-D converters operating in parallel and offset by 24 dB. These two digital signals are seamlessly recombined in the digital domain to produce a single digital output signal with a net resolution of 28 bits (Monforte, 2001). Figure 8–29A shows a view of the Solution-D microphone, and a signal flow diagram is shown at B.

B

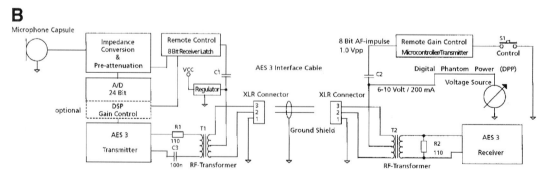

FIGURE 8–28

Details of beyerdynamic digital microphone system: view of microphone (A); signal flow diagram (B). (Data courtesy of beyerdynamic.)

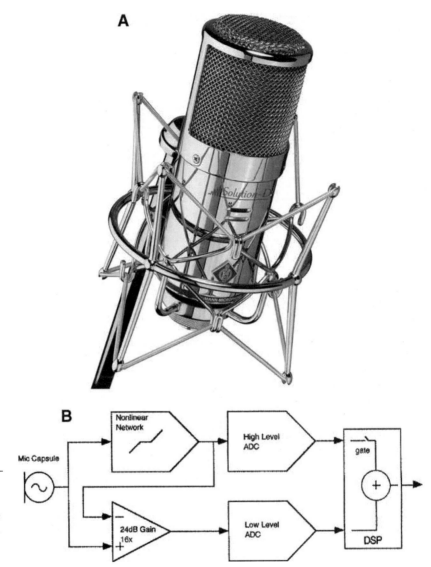

FIGURE 8-29

Details of Neumann
Solution-D digital
microphone system: view of
microphone (A); signal flow
diagram (B). (Data courtesy
of Neumann/ USA.)

Both of these microphone systems have additional digital features, including variable sampling rates, various interface formats, some degree of built-in digital signal processing, and the ability to respond to certain user commands via the digital bus. The Audio Engineering Society (AES) is actively pursuing interface standards for this new class of products.

C H A P T E R   9

OVERVIEW OF WIRELESS MICROPHONE TECHNOLOGY

INTRODUCTION

Outside of recording and broadcast studios, the wireless microphone is virtually indispensable. The technology dates back to the 1960s, and great strides in performance quality and overall reliability have been made since that time. Television, staged music and drama performances, religious services, and public meetings all make use of wireless microphones; and the ultimate freedom of movement offered by the technology is seen as a boon by everyone involved in live performance.

The earliest wireless microphones employed the commercial FM (frequency modulation) band at very low output power, and consumer FM tuners were used for signal recovery. In time, the Federal Communications Commission (FCC) allocated specific frequency bands in the VHF and UHF television ranges for wireless microphones as well as for other short-range communications needs. Today, virtually all major microphone manufacturers offer wireless systems, and the user has much to choose from. Most of these manufacturers have published detailed user guides that cover both technical matters as well as usage recommendations.

In this chapter we discuss in detail the technology involved in wireless microphone design and its application in a number of performance areas.

CURRENT TECHNOLOGY

FREQUENCY ALLOCATION

Wireless microphones can be licensed for operation in the following radio frequency (RF) ranges:

1. VHF (very-high frequency) range:
   Low band: 49–108 MHz
   High band: 169–216 MHz

2. UHF (ultra-high frequency) range:
   Low band: 450–806 MHz
   High band: 900–952 MHz

The FCC gives priority to primary users, such as TV, radio, and commercial communications activities such as cellular telephones, pagers, and two-way radio applications. Wireless microphone application is considered a secondary activity, and as such is not allowed to interfere with primary activities. On the other hand, users of wireless microphones are often subjected to interference from primary users and must find their own solutions to these problems through the choice of more appropriate, trouble-free operating frequencies.

In the US, a manufacturer or distributor of wireless microphone systems must obtain a license from the FCC to sell the equipment, but it is the responsibility of the final user/purchaser to observe and follow all recommendations regarding proper use of the equipment. The allocation of specific frequency bands varies widely around the world and within given countries, and the user must rely on good advice from both manufacturer and dealer in purchasing the correct equipment for a given application in a specified location.

Regarding overall quality of performance, the UHF range is preferred in that it provides excellent line-of-sight reception within a given venue with minimum radiation beyond the normal architectural confines of that venue.

In the future, wireless microphone operations will have to coexist with the demands of digital television and expansion of other classes of communication. Both manufacturers, specifiers and users of wireless products must keep abreast of all developments and restrictions in these areas. Figure 9–1 shows some typical international frequency allotments

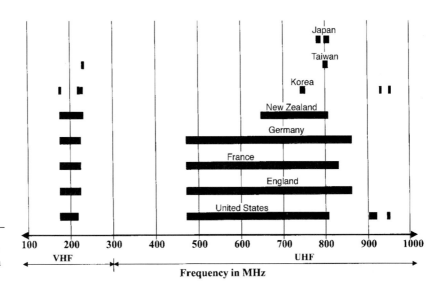

FIGURE 9–1

Some international wireless frequency allotments. (Data courtesy of Shure Inc.)

for wireless microphone activity (Vear, 2003). A more comprehensive listing is given in Ballou (2002).

TRANSMISSION PRINCIPLES

Wireless microphones work on the principle of FM transmission, as shown in Figure 9–2A. A carrier at the assigned VHF or UHF frequency is modulated by the audio program and sent to an RF amplifier contained in the microphone case itself or in a "body pack" worn by the user. The RF signal is propagated through space by a transmitter operating in the power range of 10–50 mW. For UHF transmission the radiating antenna can be housed in the lower portion of the microphone case. The antenna for the bodypack is usually in the form of a short wire that hangs externally from the case.

Signal reception is shown in Figure 9–2B. Here, the RF signal is amplified and amplitude-limited to reduce static interference; the recovered

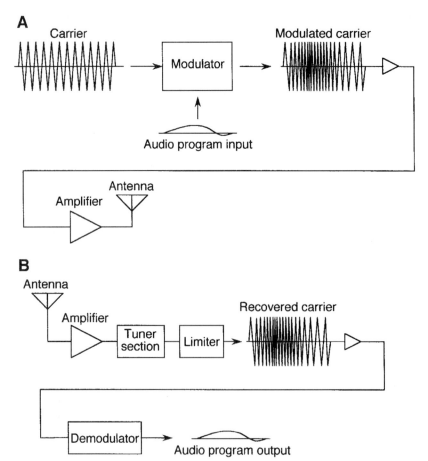

FIGURE 9–2

Principle of FM transmission and reception: simplified block diagram of a transmitter (A); simplified block diagram of a receiver (B).

carrier is then demodulated to produce audio output at the receiving end of the system.

The simple system shown in Figure 9–1 has two major problems: the potential loss of signal due to multiple reflections from transmitter to receiver as the performer moves about the stage, and the susceptibility to noise due to the low radiated power and low modulation index of the transmitter as mandated by FCC standards and the requirements for minimal interference with adjacent channels.

Diversity reception is used to overcome the multiple reflection problem, and a combination of *signal companding* (complementary compression and expansion) with *pre-* and *post-equalization* is used to reduce noise.

DIVERSITY RECEPTION

With a single receiving antenna there exists the possibility that primary and reflected carrier signals will arrive at the receiving antenna out-of-phase, causing a momentary dropout of the carrier, as shown in Figure 9–3A. This can be substantially avoided through the use of diversity reception. In diversity reception there are two receiving antennas which are spaced by a distance between one-quarter the carrier wavelength and one carrier wavelength, and this reduces the likelihood of signal cancellation. The equation relating radiated wavelength to frequency is:

$$\text{wavelength } (\lambda) = c/f \tag{9.1}$$

where λ (Greek letter *lambda*) is wavelength, c is the speed of radio waves (300,000,000 meters per second), and f is the frequency in Hz.

In the frequency range of 900 MHz, carrier wavelength is in the range of about 3×10^8 to 9×10^8 meters (about 13 in), so an antenna spacing of about 8–35 cm (3–13 in) can be used to substantially reduce carrier cancellation at the antennas. This is shown in Figure 9–3B, with the two receiving antennas (labeled A and B in the figure). Most modern diversity receivers are of the so-called "true diversity" type, in which the signals picked up by both antennas are simultaneously demodulated by two separate receivers so that the switching function from one to the other, depending on signal strength, is virtually instantaneous and free of any switching noises.

SIGNAL COMPRESSION AND EXPANSION

To reduce the basic noise floor of the relatively narrow deviation FM transmission system used with wireless microphones, complementary signal compression and expansion, known as *compansion*, are used. Figure 9–4 shows, in effect, how this works. The input signal varies over a range of 80 dB and is compressed in the transmitting circuitry to occupy a range of about 40 dB. After demodulation the signal is expanded by 40 dB to restore the original dynamic range.

A

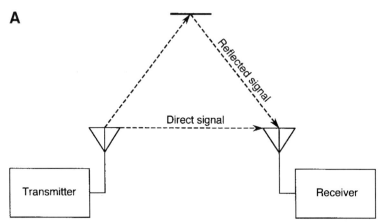

When a direct path and a reflected path are both received at the single antenna, there may be some degree of phase cancellation, resulting in weak or no output.

B

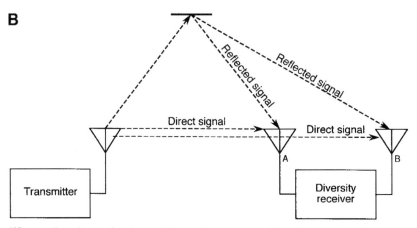

When a diversity receiver is used, two antennas, spaced by about one-fourth to one wavelength, pick up the signal, and there is a very low likelihood that cancellation will take place at both antennas simultaneously.

FIGURE 9-3

Diversity reception; single receiving antenna (A); dual (diversity) receiving antenna reception (B).

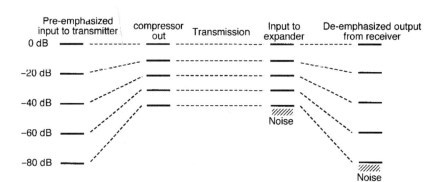

FIGURE 9-4

Principle of operation of signal companding (compression-expansion).

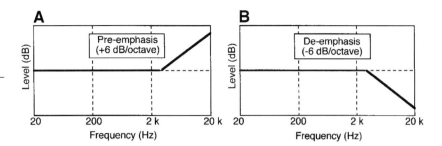

FIGURE 9–5

Typical HF pre-emphasis (A) and de-emphasis (B) curves used with wireless microphones.

Companding is normally coupled with complementary signal pre-emphasis and de-emphasis in order to minimize the audibility of noise floor modulation by the complementary gain shifts. Typical fixed pre- and de-emphasis curves are shown in Figure 9–5.

Considering the combined effects of diversity reception, companding, and complementary pre- and de-emphasis, the subjective dynamic range of a wireless microphone channel is in the range of 100 dB, considering an A-weighted noise floor as measured under no-signal conditions.

TRANSMITTER UNITS

There are two basic transmitter units: the handheld microphone and the body pack. These are shown in Figure 9–6 along with a receiving unit. The microphone transmitter is convenient for most applications, and a number of manufacturers provide a basic body on which a chosen microphone capsule assembly can be fitted.

For those applications where the microphone must be concealed (drama and musical shows), or where the user demands the freedom of both hands, the body pack is used. The body pack is worn by the user, normally in a pocket or strapped to the user's belt. A thin wire connects a standard tie-tack microphone to the body pack. Most body packs also have a line input option for use in picking up amplified instruments which may be carried around on-stage.

The transmitter's antenna wire must hang free in order to ensure proper operation. Those systems operating in the UHF high band will require only about an 8 cm (3 in) antenna, and these are often built directly into the microphone transmitter.

OPERATIONAL CONTROLS

THE TRANSMITTER

A transmitter of either type will normally have the following controls (the nomenclature may vary slightly from one manufacturer to another):

1. *Frequency selection.* This is fixed inside the transmitter and is not readily accessible to the user.

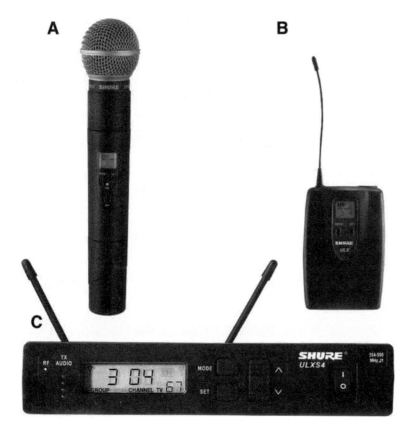

FIGURE 9-6 ————

Photograph showing a
typical wireless
microphone, bodypack
and receiver. (Photo
courtesy of Shure Inc.)

2. *Sensitivity control*. This is in effect a volume control that determines the level of the microphone signal reaching the transmitter input. It can be adjusted for each user.

3. *Power on-off*. The power should of course be switched off when the microphone is not in use in order to conserve battery life. In actual operation however, the wearer should be instructed to leave the microphone power on at all times in order to ensure program continuity.

4. *Microphone mute*. The mute switch should be engaged by the wearer only when there is a lull in use, so that normal handling noises will not be transmitted. The wearer must of course be aware of the microphone's status so that it can be turned on when needed.

5. *Status lights*:
   Microphone on/off.
   Low battery condition. Normally, a blinking of the on/off light is used to indicate low battery power.
   Signal overload. Normally, a red light will indicate the presence of too-high signal peaks, warning the user to talk at a lower level or to readjust the sensitivity control.

THE RECEIVER

Today, receivers are fairly small, usually no more than one standard rack unit high and one half-rack unit in width. While each unit can accommodate a pair of small diversity antennas, it is usually more convenient to feed all receivers from a central diversity antenna pair with a splitter/booster unit, both of which may be located at the top of the rack containing the receivers.

For each microphone or body pack in use there will of course be a corresponding receiver. It is possible to assign multiple transmitters to a single receiver, but it is obvious that only one transmitter can be used at a given time.

Receivers vary in complexity and some will have more controls and indicators than others. The following are considered essential:

1. *Frequency selection*. Must match that of the transmitter.
2. *Squelch/mute threshold control*. This control is used to set the RF sensitivity of the receiver so that it will pick up the desired local signal, but will mute when that signal is turned off.
3. *Audio level*. This control should be set for the desired output level from the receiver. It is usually set for normal line output level.
4. *Status lights*:
   Mute indicator. Indicates when the receiver is not picking up the assigned transmitter.
   RF level indicator. Indicates the level of incoming RF signal; a persistently low level indicates the possibility of signal muting.
   AF (audio frequency) level indicator. Indicates the level of audio being fed downstream.
   Diversity operation. Under normal operating conditions there should be a continuous toggling action between the two indicator lights, indicating that diversity decisions are being made more or less randomly – which is the intended operation of the system.

POWER MANAGEMENT

Since all wireless microphones and body packs are battery operated, there must be a good supply of fresh batteries available at all times. Rechargeable batteries may, over time, represent a saving over the cost of non-rechargeable batteries. However, the non-rechargeable batteries do offer the benefit longer operating time and slightly higher output voltage than the equivalent rechargeable units. Figure 9–7 shows the expected active life of several battery types. You can easily see that, in a very long evening's show, if an actor, singer, or master of ceremonies is on stage continuously for up to three hours, it will be essential to outfit the transmitter with a standard alkaline non-rechargeable battery.

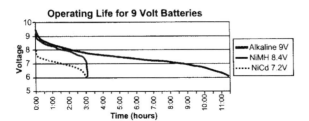

FIGURE 9–7 ————

Operating life of various
battery types. (Data
courtesy of Shure Inc.)

For the average performance lasting about three-plus hours, any rechargeable batteries that are in use will have to be replaced at some point around the middle of the show, and this may not always be convenient to do. For each rechargeable battery in use, it will be necessary to have two others on hand – one freshly charged and the other being charged. Furthermore, non-rechargeable batteries can be bought in bulk with considerable cost savings. The bottom line here is to weigh the pros and cons of rechargeable versus non-rechargeable batteries very carefully.

It is essential that one person be placed firmly in charge of all wireless microphone operations in a given application involving more than a few microphones. The major responsibility is to ensure that there will always be a supply of freshly recharged batteries and that there be a rigorous schedule for all users to have batteries replaced when needed as the show goes on.

On a more global basis, the wireless specialist must also be aware of what channels are actually available for use in all geographical venues in which a road show may be scheduled. Also, the wireless specialist must be aware of any other shows or events that may be scheduled in adjacent show or sports facilities. Not all problems can be encountered and corrected during an afternoon dress rehearsal. Patterns of RF interference often change in the evening and must be anticipated.

PERFORMANCE ENVIRONMENTS

Essentially, wireless microphones rely on line-of-sight between transmitter and receiver. The RF signal will travel easily through curtains and thin wallboard materials provided that there is no metal mesh present. Ordinarily, a wireless microphone should not be used within a distance of 3 m (10 ft) of a receiver, due to the possibility of overloading the receiver's first RF amplification stage.

Normal usage may extend up to 300 m (1000 ft) if the venue is relatively free of RF interference. In very large venues it may be advantageous to have several transmitter/receiver groups, just to make sure that excessively long distances will not be encountered. Likewise, it is important to ensure that all "roving" microphones will perform well when tested over their entire anticipated range of movement. Do not hesitate to reassign frequencies if transmission problems are encountered. Often, moving the receiver rack a small distance will cure a transmission problem. While the actual number may vary, it is generally possible to use up to 30 wireless microphones simultaneously in a given venue. In fixed installations remember that wireless microphones should be used only where there is a need for user mobility. Microphones used solely at pulpits, podiums, and lecterns should all remain wired.

FINAL COMMENTS

Do not treat the specification and application of wireless microphones as an afterthought or as a trivial matter; there is too much that can go wrong! Never make the mistake of believing that you can learn all you need to know about these systems in one simple training session. We strongly recommend that you acquire the wireless usage manuals available from many manufacturers. In particular we recommend the comprehensive guide written by Tim Vear (2003) for Shure Incorporated.

C H A P T E R   1 0

MICROPHONE ACCESSORIES

INTRODUCTION

With the exception of handheld applications, microphones are rarely used without accessories of one kind or another. Most accessories are used in mounting the microphone in a stable configuration, ranging from desk stands and flush mounts to large microphone stands or booms. Hanging mounts are often used where stands are not appropriate. Under conditions of wind or mechanical instability, wind screens and shockmounts are essential. Numerous electrical expedients, such as padding (electrical attenuation), polarity inversion, and the like may be conveniently met with in-line adapters. Likewise, stand-alone phantom power supplies for capacitor microphones are often useful. In this chapter we will cover the entire range of accessories.

THE DESK STAND

The once familiar desk stand, shown in Figure 10–1A, has largely given way to flush-mounted microphones in the boardroom and to the ubiquitous telephone handset in modern transportation terminals. Many desk stands today are considered too obtrusive and have been replaced by models with a thin gooseneck extension and small capsule. This type of assembly (Figure 10–1B) can be either mounted in a heavy base or permanently mounted on the working surface of a lectern. In either case, the microphone itself can be easily positioned for the talker.

Purely for the sake of nostalgia, some late-night TV talk show hosts may have a vintage microphone sitting on their desk. Look carefully; you will note that the performer is usually wearing a lapel microphone!

FLUSH-MOUNTED MICROPHONES

The "mike mouse," shown in Figure 10–2, is made of open cell foam and is a convenient way to keep a small microphone positioned close to

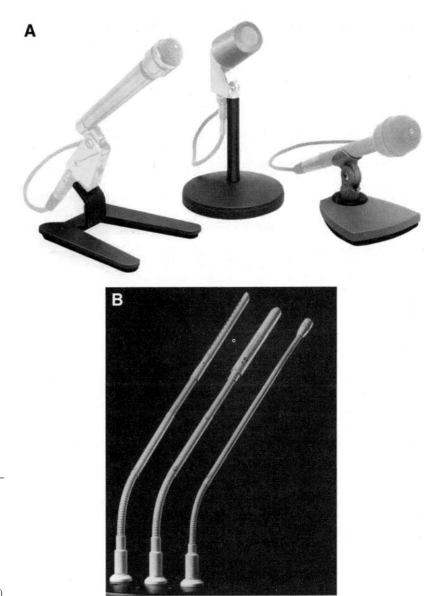

FIGURE 10–1

Microphone desk mounts:
standard models (A);
dedicated models with
slender gooseneck and
microphone sections (B).
(Photo A courtesy of
Electro-Voice; Photo B
courtesy of AKG Acoustics.)

a surface in a stable position. These used to be familiar items in board
rooms and on the stage floor in performance venues. The microphone's
position at a relatively large boundary results in minimal reinforcements
and cancellations due to reflections from the floor surface to the microphone. Today, we are more likely to see a very low profile boundary layer
(BL) microphone for such applications as these.

FIGURE 10-2

Foam floor-mount adapter,
top and bottom views.
(Photo courtesy of
Electro-Voice.)

MICROPHONE STANDS AND BOOMS

Microphone stands come in all sizes and configurations and can span adjustable heights from about 0.5 meter to 5 meters (1.5 ft to 15 ft), as shown in Figure 10–3A. The high stands require a large footprint for safety, and take great care when using them with large format microphone models. These stands are useful where sound pickup in front of and above an ensemble is sufficient. Where there is a need to position the microphone within an ensemble, then a horizontal boom assembly attached to the stand will allow positioning the microphone over the players. A comfortably large footprint is necessary, as is an appropriate counterweight to balance the microphone. Such booms are a mainstay in the popular studio.

Do not stint on quality. Good stands and booms are not cheap, and well-made models will last indefinitely. A number of recording engineers who are in the remote recording business have taken a page from those in the motion picture trades. They have gone to grip manufacturers (suppliers to the motion picture trade for stands, props, reflectors, scrims, and the like) for purpose-built microphone stands at relatively reasonable prices.

For motion picture and video field sound pickup, telescoping hand-held booms, as shown at B, are typical. Typical usage of a boom is shown in Figure 10–3C. A good operator skillfully keeps the boom-microphone assembly out of the picture, closely following the talkers and keeping the microphone pointed at them.

STEREO MOUNTS

A stereo mount is a piece of hardware that will accommodate two microphones on a single stand, positioning them securely for stereo pickup. The model shown in Figure 10–4A is articulated so that both microphone angle and spacing can be adjusted. The assembly shown at

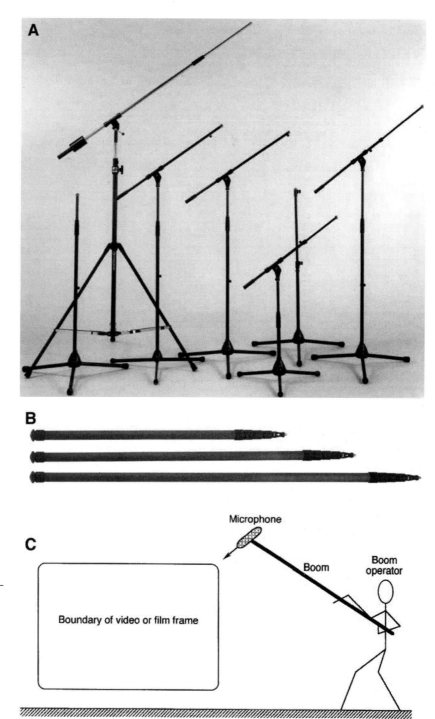

FIGURE 10-3

Typical microphone stands
and boom assemblies (A);
telescoping handheld booms
as used in film and video
work (B); boom in
operation (C). (Photo A
courtesy of AKG Acoustics;
Photo B courtesy of M.
Klemme Technology.)

Microphone

Boom

Boom
operator

Boundary of video or film frame

A

B

C

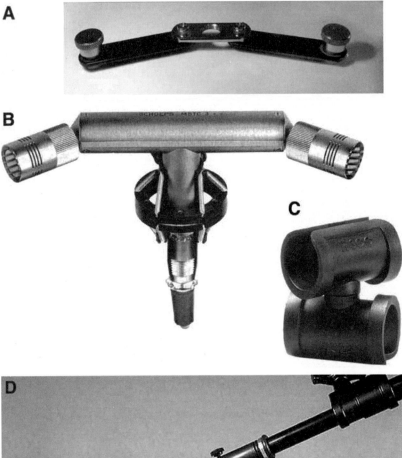

D

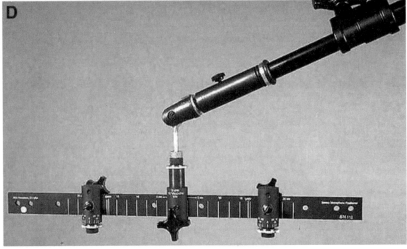

E

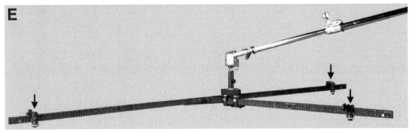

FIGURE 10-4

Stereo mounts; articulated type (A); dedicated ORTF assembly (B); closely spaced, rotatable assembly (C); long adjustable stereo bar (D); and Decca tree (E). (Photo B courtesy of Schoeps GmbH; Photo C courtesy of AKG Acoustics; Photos D and E courtesy of Audio Engineering Associates.)

B provides ORTF mounting for Schoeps cardioid capsules, and the assembly shown at C provides XY or MS pickup using small format capacitor microphones. The horizontal bar shown at D may be used for a wide range of spacing between the stereo microphone pair. A so-called Decca tree is shown at E and is used in front of large orchestral groups in the manner of the British Decca record company. Small arrows indicate the positions of the three microphones.

PERMANENT MICROPHONE INSTALLATIONS

Music schools and festival sites have a need for quasi-permanent microphone mounting so that ensemble and recording changes on-stage can be made quickly. While there will always be a requirement for floor stands as needed, overhead requirements can normally be met by a system of breast lines, pulleys, and cables attached to relatively permanent microphone receptacles. Most of these systems have been improvised and fine-tuned over time. Figure 10–5 shows details of a three-way

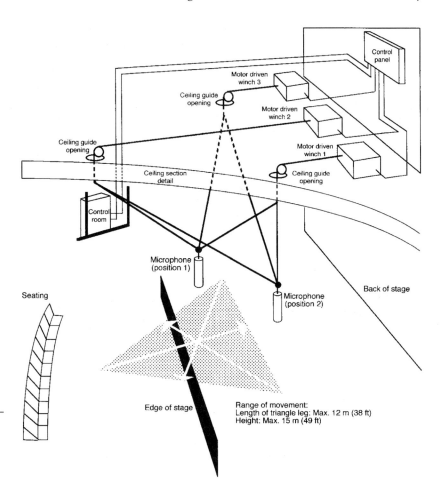

FIGURE 10–5

Details of a three-point winch system for positioning a microphone. (Data after EMT.)

winch system designed some years ago by the German EMT company. This figure will give the enterprising designer an idea of the complexity involved. Such a system as shown here would be useful with a remotely adjustable stereo microphone and is ideal for use with the Soundfield microphone.

CABLE MOUNTS

For many performance applications, one or more single hanging microphones may be used with the arrangement shown in Figure 10–6A. This mount uses a clamping action on the cable, providing a service loop of cable that permits the microphone to be angled as needed. A horizontal breast line can be tied to the cable loop for added positioning and minimization of cable twisting over time.

The method shown at B relies on a flexible wire to position a small electret microphone at the desired angle. An array of such hanging microphones would be effective for choral pickup in a house of worship.

WIND AND POP SCREENS

The nature of wind noise is shown in Figure 10–7. A puff of wind travels perhaps 1.5 m/s (5 ft/sec), while sound travels at a speed of 344 m/s (1130 ft/sec). Furthermore, the motion of wind occurs along a narrow path straight ahead of the talker while speech radiates relatively uniformly over the forward hemisphere.

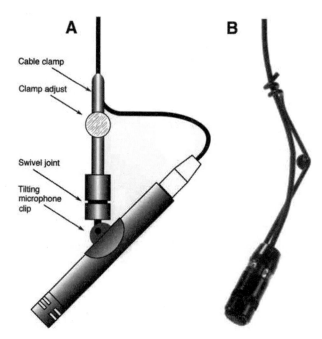

A

Cable clamp

Clamp adjust

Swivel joint

Tilting microphone clip

B

FIGURE 10–6

Cable mounts; a cable clamp for swiveling and tilting of microphone (A); flexible wire for aiming small microphone (B). (Photo B courtesy of Crown International.)

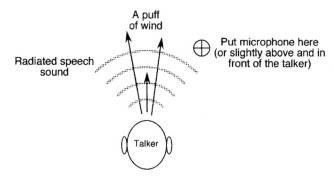

FIGURE 10-7

Action of wind and speech on a microphone; the nature of wind noise.

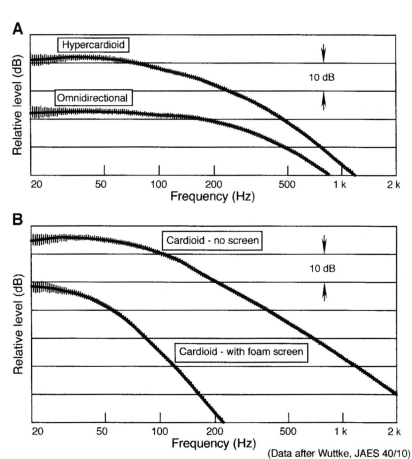

FIGURE 10-8

Relative effects of wind on noise generation in omnidirectional and hypercardioid microphones (A); noise generation in a cardioid microphone with and without a foam screen (B). (Data after Wuttke, 1992.)

Most users not familiar with microphones feel that they must talk directly into the microphone; the important thing is that the microphone be pointed at their mouth, and that is best done if the microphone is placed to one side or slightly above the talker along the centerline and aimed at the talker's mouth.

Figure 10–8 shows some spectral details of wind noise. For a fixed wind velocity, the data at A shows that the difference in the noise generated by a hypercardioid microphone is about 20 dB higher than with an omnidirectional microphone. The data shown at B shows the performance of a cardioid microphone with and without a foam windscreen. Note that the effectiveness of the windscreen diminishes at low frequencies. In order to maintain the high midband reduction in wind noise, the size of the wind screen has to be increased substantially.

Windscreens can be problematic. They nearly always affect frequency response, and if they are too dense they will affect directionality as well. Wuttke (1992) points out that any windscreen used with a gradient microphone should have a clear inner open cavity in the region of the microphone element so that the gradient effect can establish itself directly at the microphone's transducing element.

Wind effects do not only occur outdoors. The author has encountered several concert halls in which high volume velocity, low particle velocity air handling systems produced wind noise in gradient microphones positioned 3–4 m (10–13 ft) above the floor.

A typical foam windscreen for handheld applications is shown in Figure 10–9A. An oblong mobile screen is shown at B, and a so-called

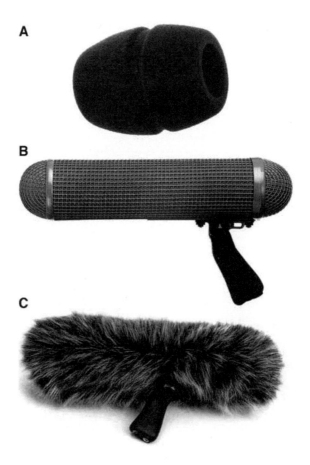

FIGURE 10-9 ———

Windscreens; a typical foam screen for handheld or studio use (A); handheld screen for field use (B); "furry shroud" for field use (C). (Photo A courtesy of AKG Acoustics; Photo B and C courtesy of beyerdymamic.)

"furry shroud" is shown at C. These assemblies are intended for field use in television and motion picture recording and are normally used with rifle microphones mounted on booms or poles.

THE POP SCREEN

A nylon pop screen is shown in typical studio usage in Figure 10–10. It is a virtually perfect antidote to the "popping" sounds of *b* and *p* from vocalists and announcers in studio applications. It is semitransparent, so that the artist can easily see the microphone, and it has virtually no attenuation at high frequencies as it effectively disperses any puff of wind in the direction of the microphone. It is normally attached to microphone hardware by a small clamp or by a flexible wire, as shown.

FIGURE 10-10 ————

A nylon pop screen in normal operating position. (Photo courtesy of Shure Inc.)

SHOCK MOUNTS

Significant floor vibration transmission through the microphone stand is not often a problem, but stands may be inadvertently hit by studio performers. When this occur some form of shock mounting is necessary. Figure 10–11 shows the transmission curve for a typical vibration isolation system. Note that below the system resonance frequency, f_n, the vibration transmission coefficient is constant. Above the resonance frequency the transmission coefficient falls steadily to a low value. In practice, we would prefer f_n to be low enough so that negligible audible vibration will be transmitted to the microphone. In order to do this we need a vibration isolator, commonly called a shock mount.

A typical assembly is shown in Figure 10–12. A normal isolation tuning frequency for a large format microphone might be in the 8 to 12 Hz range, well below the normal hearing range. Note that the microphone is entirely suspended by the elastic bands and is not in direct contact with the frame. Note also the microphone cable hanging from the bottom of the microphone. If for any reason the microphone cable is pulled taut and fastened to the stand, then the effect of the entire shock mount may be negated. There should always be a loose service loop of cable as it exits the microphone, and that loop should be taped to the microphone stand or boom to avoid any downward gravity loading on the microphone due to excessive cable length.

Microphones and their shock mounts are designed as a system. It may be possible to mix and match models to some degree, but if the resonance varies substantially the effectiveness of the shock mount may be compromised.

Foam rubber isolating rings are routinely used in microphone mounts that are sold with microphones. Because of their relatively low mechanical compliance, these devices account for little in the way of effective vibration isolation.

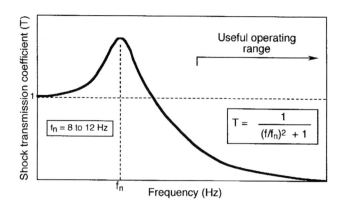

FIGURE 10-11

Universal vibration transmissibility curve through a shock mounting system.

FIGURE 10-12 —————

A typical large format capacitor microphone and its associated shock mount. (Photo courtesy of Neumann/USA.)

IN-LINE ELECTRICAL ACCESSORIES

Figure 10–13 shows a number of in-line passive devices that can be used in microphone circuits. The assemblies shown at *A* are known as "turnarounds" and may be used when, for whatever reasons of miswriting or design, XLR cables do not mate. The assemblies shown at B are polarity or phase inverters and are used to correct polarity mistakes in wiring. Some older line-level electronic items have inverted outputs that may require correction. Polarity inverters may be used at microphone level with phantom powering, but cannot be used with T-powering.

An in-line balanced loss pad is shown in Figure 10–13C. Loss values may be in the range of 20 dB or greater. The assembly may be used with dynamic microphones, but not with any kind of remote

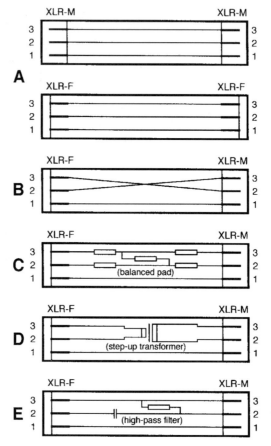

FIGURE 10-13

Various electrical
accessories; male-female
"turnarounds" (A); polarity
inverter (B); loss pad (C);
low-to-high impedance
transformer (D); and
low-pass filter (E).

powering for capacitor microphones. An in-line transformer, shown at
D, can be used to match a low impedance dynamic microphone with a
high impedance input in some nonprofessional applications. The in-line
LF-cut equalizer shown at E likewise may be used in nonprofessional
applications to correct for proximity effect in dynamic cardioid microphones. The microphone splitter was discussed in Chapter 8, under
Microphone Splitters.

AUXILIARY MICROPHONE POWER SUPPLIES

Details of a phantom power supply are shown in Figure 10–14. This unit
enables a P-48 microphone to be used with a small PA-type mixer that
may not have built-in phantom powering.

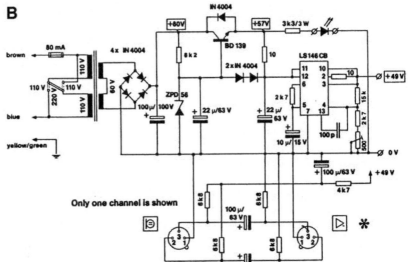

FIGURE 10-14

Photo (A) and circuit diagram (B) for AKG Acoustics N62E auxiliary phantom power supply. (Figure courtesy of AKG Acoustics.)

C H A P T E R   1 1

BASIC STEREOPHONIC
RECORDING TECHNIQUES

INTRODUCTION

Modern stereophonic recording, or stereo as we normally call it, makes use of many diverse microphone arrays and techniques. At the basis of them all are a set of fundamental two- or three-microphone arrays for picking up a stereo sound stage for reproduction over a pair of loud-speakers. In stereo reproduction the listener is able to perceive images on the stereo sound stage which may span the entire angular width of the loudspeaker array. The sound sources that are perceived between the loudspeakers are known as "phantom images," because they appear at positions where there are no physical, or real sources of sound.

HOW STEREO WORKS: A SIMPLIFIED ANALYSIS
OF PHANTOM IMAGES

It is essential that the reader has an accurate, if only intuitive, knowledge of how phantom images are formed. In Figure 11–1 we show how a set of real sound sources located progressively from the center to the right-front of the listener produce time-related signals (phasors) at the ears. At frequencies below about 700 Hz, each single sound source (S_1 through S_5) produces a signal at the right ear that leads the slightly delayed signal at the left ear. The ears detect this as *a phase difference* at low frequencies. At high frequencies (above about 2 kHz), the ears rely primarily on the amplitude differences at the two ears, and a louder signal will appear at the right ear due to the shadowing effect of the head. These two sets of cues reinforce each other, and the ear–brain combination interprets the differences at the ears as a set of apparent sources spanning the front to front right.

For a sound source (S_1) located directly in front of the listener, the low frequency phasors will be equal at both ears, as will be the shadowing at

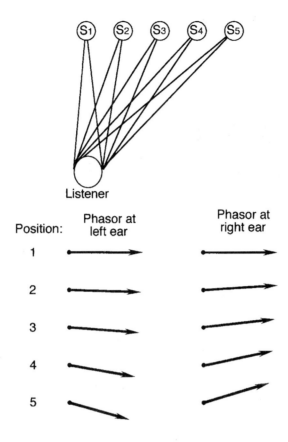

FIGURE 11-1

Phasor analysis for real sources with positions ranging from center to right-front of the listener.

high frequencies. These cues reinforce the judgment that the sound source is directly ahead of the listener.

With only two loudspeakers not exceeding a spread of about 60 degrees, we can create low frequency phasors for sound sources that span the entire listening angle between the loudspeakers. To create the phasors for a sound source directly ahead of the listener, all we need to do is feed the same signal to both loudspeakers, as shown in Figure 11–2. Here, each ear receives two signals, one slightly delayed with respect to the other. Since the program content of the signals is the same, we can combine their phasors at each ear to create new phasors, shown here as L_T and R_T. Since these two phasors are equal (that is, they have the same amplitude and phase relationships), the ear–brain combination will interpret them as producing a sound source directly in front of the listener – as long as the listener is facing forward.

By varying the amounts of left and right components, phasors can be created that will simulate all positions between the left and right loudspeakers. As we will soon see, a single pair of microphones can easily create this effect across the left-to-right span of the playback loudspeaker array.

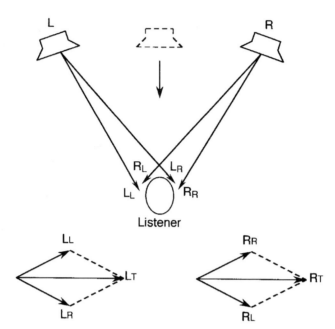

FIGURE 11–2

Two loudspeakers creating equal net phasors at both ears.

THE EFFECT OF DELAYED SIGNALS AT HIGH FREQUENCIES

In addition to the effect of low frequency phasors, small discrete signal delays between the two microphones at high frequencies can also influence phantom source localization. In normal recording operations, both amplitude and time delay cues are at work, and their combined effect on phantom source localization can be roughly estimated from the data presented in Figure 11–3. This data was presented by Franssen (1963) and gives reasonably accurate estimated stereo stage localization for wideband signals such as the spoken voice. An example showing use of the graph is given in Figure 11–3.

In subsequent discussions in this chapter we will refer to approximate localization analyses as determined by Franssen's method. These will be shown as horizontal plots with specific indications of left, left-center, center, right-center, and right images as determined from the data in Figure 11–3.

COINCIDENT STEREO MICROPHONE ARRAYS

Coincident microphone arrays consist of a pair of directional microphones located virtually one atop the other and both individually adjustable in their lateral pickup angles. As such, they respond only to amplitude cues in the program pickup, since their proximity precludes any time related cues. Coincident arrays are often referred to as X-Y arrays.

Tradeoffs between amplitude and delay in determining approximate localization on the stereo soundstage:

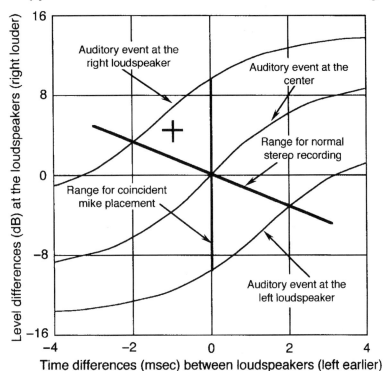

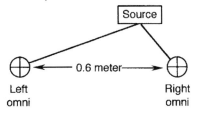

Example:

Source leads in right channel by 1 millisecond.

Source is 5 dB louder in right channel than in the left.

FIGURE 11-3

Franssen's data for determining wide-band stereo localization with both amplitude and time cues.

Where will the stereo image appear?

1. On horizontal axis, go to -1 (left channel lags).
2. On vertical axis, go to +5.
3. Intersection is shown in figure.

CROSSED FIGURE-8s: THE BLUMLEIN ARRAY

The most notable example of the coincident array is a pair of figure-8 microphones arrayed at a lateral angle of 90°, as shown in Figure 11–4A. This array, first described by Blumlein (1931, reprinted 1958), is unique in that the sine and cosine angular relationships between the two pickup pattern lobes retain constant signal power over the pickup angle due to the relationship:

$$(\sin \theta)^2 + (\cos \theta)^2 = 1 \qquad (11.1)$$

This identity ensures that any sound source located in the front quadrant of the array will be picked up with uniform power, since acoustical power is proportional to the square of its respective components. Specifically, when a sound source is directly in the middle of the array, the values of both sin 45° and cos 45° will be 0.707, representing a −3 dB reduction in level at both left and right loudspeakers. These two half-power quantities will add acoustically in the listening space, yielding a power summation of unity. Localization of the Blumlein array is shown as determined by the Franssen graph shown in Figure 11–4B.

Any discussion of the Blumlein array leads directly to a discussion of the *panpot* (panoramic potentiometer), a dual fader arrangement that has one signal input and two signal outputs. Circuit details are shown in Figure 11–5A, and the two outputs are as shown at *B*. Panpots are found in the microphone input sections of virtually all recording consoles and can be used to steer a given input signal to any position on the two-channel stereo sound stage. The two fader sections give sine and cosine values corresponding to the rotary fader setting, providing uniform acoustical power whatever the setting. In this regard the panpot is analogous to the Blumlein crossed figure-8 microphone pair operating in its front or back quadrant.

The crossed figure-8 array maps the frontal quadrant of its pickup with virtual accuracy. Likewise, the back quadrant is equally well picked up, but in opposite polarity to that of the front quadrant. The left and right side quadrants are picked up with their two components in reverse polarity (often described as *out of phase* or in *antiphase*), and this creates a spatial ambiguity. As long as the side quadrants and the back quadrant are relegated to "back of the house" reverberation, there should be no adverse problems due to the polarity reversal.

The crossed figure-8 array is also known by the term *Stereosonic* (Clark et al., 1958) as it was modified and applied by the Electrical and Musical Industries (EMI) during the early days of the stereo LP in England. In their application of the Blumlein array for commercial recording, EMI recording engineers always introduced a slight amount of in-phase crosstalk above about 700 Hz between the stereo channels. The technique, known informally as "shuffling," was used to match phantom image localization at high frequencies with that normally produced by low frequency phasor reconstruction at the listener's ears (Clark et al., 1958).

In summary, Blumlein array performance:

1. Produces excellent stereo stage lateral imaging, due to the self "panning" aspect of the frontal sine and cosine pickup lobes. Image analysis by means of Franssen's data indicates that the primary sound sources will fill the entire stereo stage from left to right.

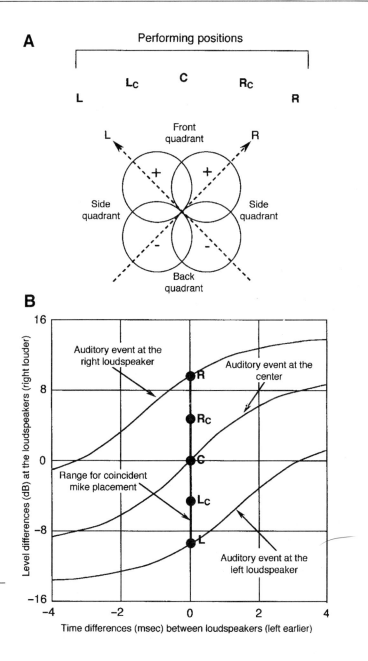

FIGURE 11-4

Crossed figure-8s (A); localization via Franssen (B).

2. Conveys an excellent sense of acoustical space, due to the added pickup of reflected and reverberant signals in the recording space.

3. Often presents difficulties in microphone placement; specifically, a wide array of performers may require that the microphone pair to be placed too far from the performers for the desired degree of presence.

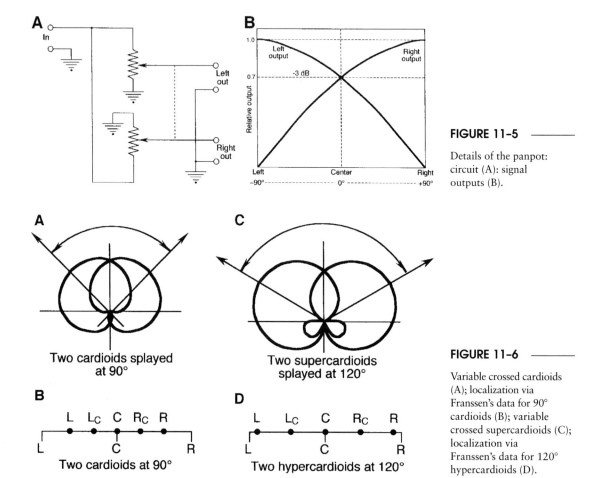

FIGURE 11-5

Details of the panpot: circuit (A): signal outputs (B).

A Two cardioids splayed at 90°

B Two cardioids at 90°

C Two supercardioids splayed at 120°

D Two hypercardioids at 120°

FIGURE 11-6

Variable crossed cardioids (A); localization via Franssen's data for 90° cardioids (B); variable crossed supercardioids (C); localization via Franssen's data for 120° hypercardioids (D).

CROSSED CARDIOID ARRAYS

Figure 11–6A shows a crossed pair of cardioids that can be splayed over an angle of 90° to 135°. When the splay angle is about 90° the sound stage appears center-heavy (i.e. mono-dominant), and sources located at far-left and right will appear panned well into the array. This is not a useful configuration, and most engineers will choose to widen the microphone splay angle up to 135°. This has the effect of correcting mid-stage balance with the sources at far left and right. But even then, the overall stereo stage may appear a bit too narrow.

A better approach is shown in Figure 11–6C, where a pair of splayed supercardioid microphones are used at an angle of 120°. Here, the relatively narrow front lobe will spread left, center, and right stage events naturally, while the smaller back lobes will pick up more room reflections and reverberation. In general, the crossed supercardioids, or hypercardioids, make a good alternative to crossed figure-8 patterns in live recording spaces, since the rear lobes will pick up less room sound and its concomitant noise. Figure 11–6B and D shows the above two orientations as they appear in the Franssen localization analysis.

In summary:

1. Crossed cardioid performance produces a center-oriented stereo stage and, as such, may be a useful adjunct in studio pickup.

2. It results in wider imaging when the splay angle is increased.

3. It has excellent monophonic compatibility (left-plus-right), due to the absence of antiphase components in the two signals.

4. Splayed supercardioids or hypercardioids are probably the most useful, offering a good compromise between wide-stage pickup and direct-to-reverberant balance.

A VARIATION ON X-Y TECHNIQUES: MID-SIDE (M-S) PICKUP

Figure 11–7A shows a typical M-S recording setup. A cardioid pattern (the *M* component) is forward facing, and a figure-8 pattern (the *S* component) is side facing. Both *M* and *S* components are often recorded separately and are combined in a matrix (sum-difference) network as shown to produce resultant left and right pickup patterns. An alternate matrix transformer circuit is shown at B.

The system offers excellent monophonic compatibility and considerable postproduction flexibility. For example, the *S* component can be reduced to produce a narrowing of the resultant array. Alternatively, the *S* component can be increased to produce a wider resultant array.

Primarily as a result of this degree of postproduction flexibility, M-S pickup has always enjoyed great favor with broadcast engineers, who have to deal with concerns of FM and television mono and stereo compatibility in a wide marketplace.

EQUIVALENCE OF MS AND XY

Any XY pair can be converted into an equivalent MS pair, and vice versa. Examples of this are shown in the data of Figure 11–8. The charts shown at B and C can be used to define the MS form for any given XY pair.

While a cardioid pattern is normally used for the *M* component, it is not essential; omnidirectional or hypercardioid patterns can be used as well if the resultant left and right patterns and their angular orientation fit the technical needs of the engineer. Those readers who wish to study M-S recording in detail are referred to the work of Hibbing (1989) and Streicher and Dooley (1982, 2002).

In summary, M-S performance:

1. Has excellent stereo-to-mono compatibility for broadcasting.

2. Has flexibility in postproduction remixing.

3. Is easy to implement in the field.

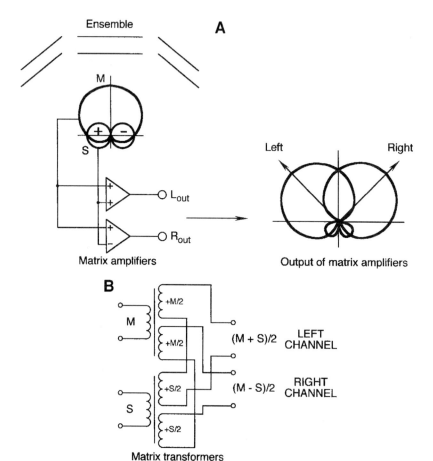

FIGURE 11-7

M-S system: a front-facing cardioid and side-facing figure-8, when mixed as shown, are equivalent to a pair of splayed cardioid microphones (A); alternate method of achieving sum and difference outputs using transformers (B).

SPACED MICROPHONE RECORDING TECHNIQUES

The beginnings of stereo recording with spaced microphones go back to the experiments carried out by Bell Telephone Laboratories (Steinberg and Snow, 1934) during the 1930s. Snow (1953) describes a wall of microphones, each communicating with a matching loudspeaker in a different space, as shown in Figure 11–9A. As a practical matter this array was reduced to the three-microphone/loudspeaker array shown at B. Later, this three-channel array gave way to an arrangement in which the center microphone was bridged (center-panned) between the left and right channels.

There is a strong tradition of using spaced omnidirectional microphones for stereo recording, especially for smaller musical forms, such as chamber music and solo piano. In Figure 11–3 we observed the stereo stage as recreated by a single pair of omnidirectional microphones separated by about 0.6 m (2 ft). When such a pair is in proximity to a small group of players, then both amplitude cues and time cues can be useful

FIGURE 11–8

Several MS forms and their equivalent XY forms (A); conversion from XY to MS (B and C); example: let an XY pair consist of a two supercardioids splayed at an angle of 120° (2α); from B, locate the supercardioid symbol and the right axis and move to the left unto the 60° curve is intersected; then, read downward to the horizontal axis and note that the required M directivity pattern will be almost halfway between supercardioid and figure-8; then, going to the same position on the horizontal axis at (C), move upward until the 60° curve is intersected; finally, move to the right axis and note that the S level will have to be raised about 2 dB relative to the M level. (Data at B and C courtesy of Sennheiser.)

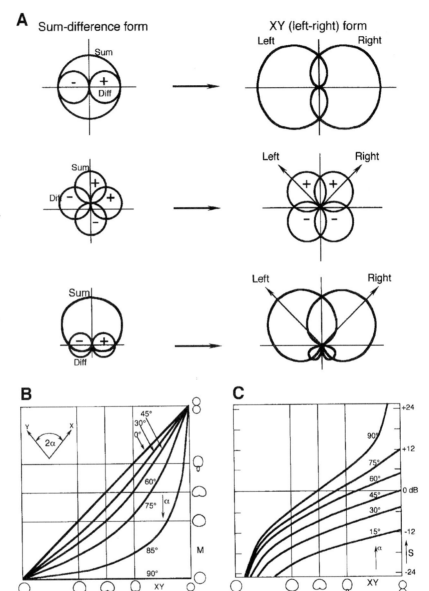

in defining the reproduced stereo stage. The images so produced may not have the localization precision or specificity of those produced by a coincident pair, and many people, engineers as well as musicians, often prefer the so-called "soft edged" quality of these recordings. The analogy of slightly blurred optical imaging is quite apt.

Actually, there is no reason why spaced cardioids should not be used – specially in large reverberant spaces. Traditionally, however, most recording engineers who prefer the imaging qualities of spaced microphones tend to favor omnidirectional models or subcardioids.

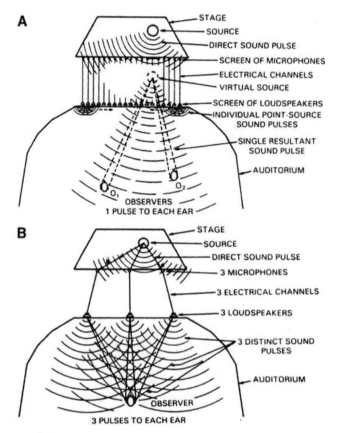

A
- STAGE
- SOURCE
- DIRECT SOUND PULSE
- SCREEN OF MICROPHONES
- ELECTRICAL CHANNELS
- VIRTUAL SOURCE
- SCREEN OF LOUDSPEAKERS
- INDIVIDUAL POINT-SOURCE SOUND PULSES
- SINGLE RESULTANT SOUND PULSE
- AUDITORIUM
- O_1 O_2
- OBSERVERS 1 PULSE TO EACH EAR

B
- STAGE
- SOURCE
- DIRECT SOUND PULSE
- 3 MICROPHONES
- 3 ELECTRICAL CHANNELS
- 3 LOUDSPEAKERS
- 3 DISTINCT SOUND PULSES
- AUDITORIUM
- OBSERVER
- 3 PULSES TO EACH EAR

FIGURE 11-9

Spaced microphones; a wall of microphones and loudspeakers (A); three microphones and three loudspeakers (B). (Figure courtesy of the Society of Motion Picture and Television Engineers.)

If two omnidirectional microphones are spaced apart more than about 1 m (3.3 ft), there may be a "hole in the middle" of the reproduced sound stage. This occurs because of reduced signal correlation between the two microphone transmission channels, and hence very little that the ears can lock onto to form center-oriented phantom images. A practical example of a two-microphone setup is shown in Figure 11–10A. Here, the performers in a string quartet are arrayed left to right, and the Franssen analysis of stereo localization is shown at B with a 0.67 meter microphone spacing. If the microphone spacing is increased to, say, 1 meter, the Franssen analysis of localization gives the results shown in Figure 11–10C. Here, the hole in the middle becomes quite evident.

For large groups such as an orchestra, it is essential that there be a center microphone panned equally between the two stereo channels. When this is done, the hole in the middle is effectively filled in, and the listener senses a continuum of sound from left to right, albeit with some loss of image specificity. Another important characteristic of such wide microphone spacing is the generation of added early time cues. As an example, consider a left-center-right microphone array spanning a total width of about 8 m (25 ft). A sound source on-stage fairly close to the right microphone will be heard first in the right loudspeaker, followed

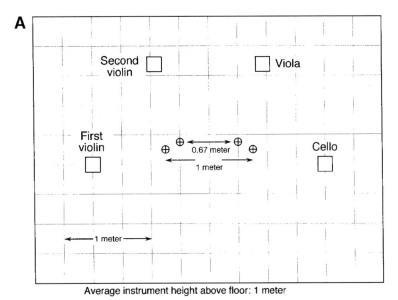

A

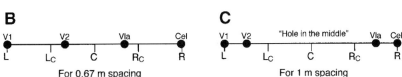

FIGURE 11-10

Two spaced omnis with a string quartet (A); Franssen analysis for 0.67 m spacing (B); Franssen analysis for 1 m spacing (C).

closely by a center phantom signal delayed approximately 12 ms and a later signal at the left loudspeaker delayed a further 12 ms. These two delayed signals will be at a reduced level, and they will effectively simulate early acoustical reflections in the recording space itself, as shown in Figure 11–11A. Their effect on the subjective impression of music may be seen in the data shown in Figure 11–11B (Barron, 1971), where they may influence the perception of tone coloration.

A typical large ensemble may be recorded as shown in Figure 11–12. The proportions are those the author has found to work in most instances. In normal acoustical spaces, omnidirectional microphones will work well, but in rooms that tend to be fairly live, subcardioids may work better because of their back rejection of about 3 dB. The level setting of the center microphone is critical; it should be at least about 4 dB lower in level with respect to the left and right microphones and should be just sufficient to avoid the sense of a hole in the middle. Any greater level of the center-panned signal is apt to make the stereo stage appear too mono-heavy.

It is worth mentioning that most American symphonic recordings made during the decades of the fifties and sixties were made basically in this manner. These recordings have always been present in commercial disc libraries, and they are as highly regarded today as they were in their heyday.

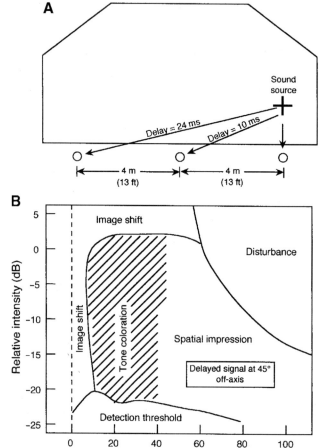

FIGURE 11–11

Widely spaced microphones
may simulate the effect of
early acoustical reflections
(A); Barron's data on the
subjective role of reflections
in a performance
environment (B).

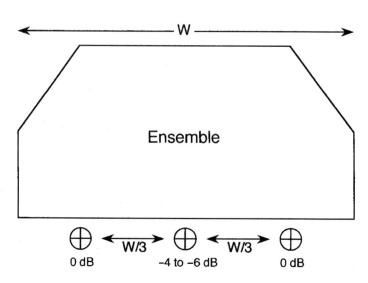

FIGURE 11–12

Three spaced omnis with
orchestra.

In summary, in spaced omnidirectional recording using two microphones:

1. Microphones are normally spaced no farther apart than about 1 m.
2. The array can be placed relatively close to a small group of performers, creating a good sense of intimacy while retaining an impression of room ambience.
3. The technique is useful primarily when precise imaging of instruments on the stereo sound stage is not essential.

In spaced omnidirectional recording using three microphones:

1. Wide spacing of microphones creates the effect of added early acoustical reflections, thus enhancing a sense of ambience.
2. The center microphone "locks" the middle of the ensemble in place, but often with relatively little precise localization.
3. Ample pickup of room reflections and reverberation retains a good sense of acoustical space, along with significant direct sound from the ensemble.

NEAR-COINCIDENT RECORDING TECHNIQUES

Near-coincident techniques employ a pair of directional microphones, closely spaced and splayed outward, and as such combine some of the attributes of both coincident and spaced microphone techniques. Some options are shown in Figure 11–13. The near-coincident microphone pairs combine both amplitude and delay effects in producing a stereo sound stage. They are useful in a wide range of acoustical settings and may be altered in both spacing and splay angle as required. Actually, there are innumerable possible configurations, and the four shown here are examples of some that have been described in the literature.

The ORTF (Office de Radio-Television Diffusion Française) technique was developed by the French Broadcasting organization, while the NOS (Nederlandsch Omroep Stichting) technique was developed by the Dutch broadcasting organization. The Faulkner and Olson (1979) Stereo-180 arrays were developed by independent recording engineers. Many engineers and musicians favor the ORTF array in particular and use it in place of the more traditional coincident cardioid array (Ceoen, 1970). Figure 11–14 shows a Franssen analysis of localization properties of the ORTF and NOS arrays.

WILLIAMS' SUMMARY OF STEREO SPACED AND SPLAYED CARDIOIDS

Since the formalization of the ORTF near-coincident approach more than 30 years ago, many engineers have experimented with their own splay angles and microphone separation, perhaps in an effort to define

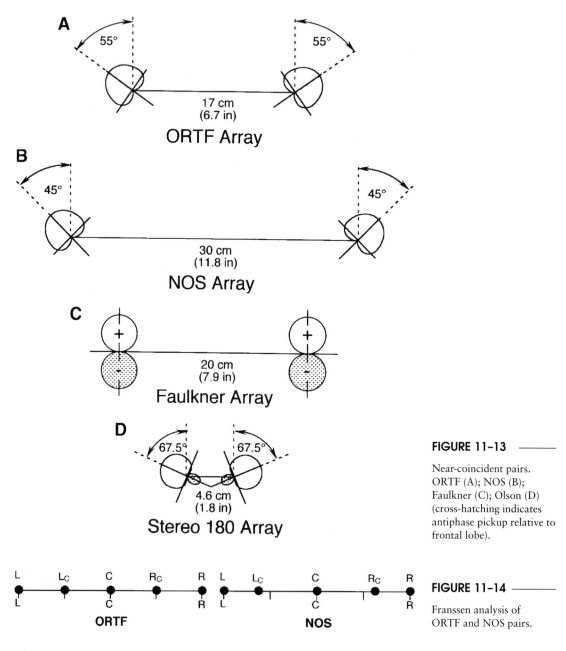

FIGURE 11-13

Near-coincident pairs.
ORTF (A); NOS (B);
Faulkner (C); Olson (D)
(cross-hatching indicates
antiphase pickup relative to
frontal lobe).

FIGURE 11-14

Franssen analysis of
ORTF and NOS pairs.

their own unique stereo signature. Williams has analyzed many microphone patterns in various splay angles and separation with the aim of
defining reasonable bounds on their performance. The data shown in
Figure 11–15 shows the useful range of cardioid microphone separation
and splay angle (Williams, 1987). The effective recording angles are indicated on the curves themselves for a variety of separation and splay angle
values. Those values in the crosshatched areas are not recommended

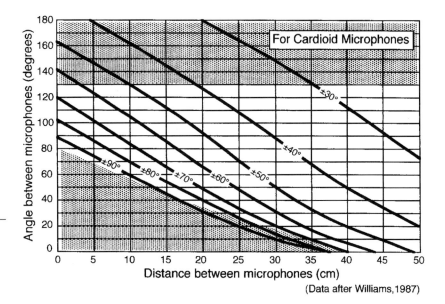

FIGURE 11-15

Williams' data showing useful combinations of separation and splay angle for a cardioid microphone pair.

because of problems in stereo imaging, ranging from insufficient stereo separation to a tendency for separation to be polarized at the loudspeakers.

In summary, near-coincident microphone performance:

1. Combines the image specificity of coincident arrays with the enhanced sense of image spatiality imparted by spaced microphones.
2. Allows considerable leeway in choice of splay angle and microphone spacing for modifying the recorded perspective.

MIXED STEREO MICROPHONE ARRAYS

Most engineers who make commercial recordings use mixed arrays in which a central pair, either coincident or near coincident, is combined with a flanking omnidirectional pair. The combination provides excellent flexibility and allows the engineer to alter perspectives in the orchestra without necessarily making adjustments in the microphone positions themselves.

For example, a shift from a close-in perspective to a somewhat more distant perspective can be made merely by altering the amount of the spaced pair in the overall mix. Care must be taken not to exceed a level range here of perhaps ±1.5 to 2 dB, otherwise the changes may be too apparent. Of course it goes without saying that such changes, if of a running nature throughout a large work, must be musically pertinent in the first place.

Some of the more often used setups of this type are shown in Figure 11–16. The ORTF-plus-flanking-omnis setup is shown at A. As a matter of taste, some engineers may replace the ORTF pair with a pair of widely splayed (120°) hypercardioids, and the omnis may be replaced with subcardioids in especially live rooms.

The variation shown at B makes use of four subcardioids. The center pair may be spaced up to 0.5 meter (20 in), and the splay angle between them set in the range of 120°. This mixed array provides good flexibility, but overall it tends to pick up considerable room ambience and reverberation.

The so-called "Decca tree" is shown at C. Developed by the English Decca Record Company in the 1950s, the array makes use of three identical microphones on a "tree" in the center, with the middle microphone gain set slightly lower than the left and right microphones. Flanking

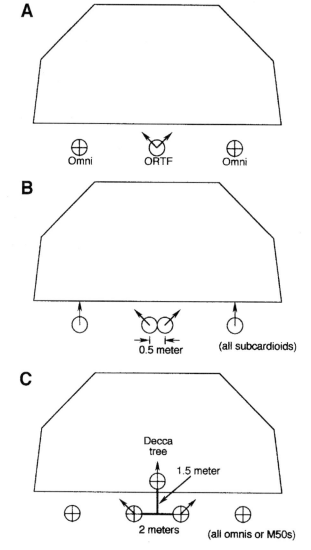

FIGURE 11-16

Mixed arrays: ORTF plus flanking omnidirectional microphones (A); four subcardioids (B); the Decca tree (C).

microphones of the same type are also used. The microphone of preference here has been the Neumann M-50, which was discussed in Chapter 3. The M-50 is essentially an omni microphone at low frequencies, exhibiting a rising HF response above about 1 kHz and reaching a level 6 dB higher than at low frequencies. Additionally, the M-50 tends to become more directional as its on-axis response increases. The array produces a sound rich in room ambience, along with an abundance of detail at high frequencies. The only difficulty associated with use of the Decca tree is suspending the three central microphones in a stable manner. A little care is needed, along with a fairly robust microphone stand.

STEREO LISTENING CONDITIONS

The width of the stereo included listening angle is a subject of considerable discussion. Generally, stereo localization is at its most stable when the loudspeaker–listener angle is moderate. At the same time, however, the angle must be large enough to convey a realistic performance stage. Most professional recording engineers would agree that an included listening angle not exceeding about 45° or 50° is about optimum.

It goes without saying that the listening environment should be fairly symmetrical about the centerline of the loudspeaker array and that the listening space be free of noticeable flutter echoes and strong standing waves. Regarding frequency response of the loudspeakers as they react with the listening space, the response should be fairly uniform over the range from about 160 Hz to about 4 kHz. Below 160 Hz a broad rise of about 2 or 2.5 dB may enhance the presentation, especially at moderate listening levels. Above about 4 kHz many listeners prefer a uniform rolloff of the response slightly (no more than 2 or 3 dB). Above all, use the best loudspeakers that your budget can maintain.

C H A P T E R   1 2

STEREO MICROPHONES

INTRODUCTION

For many applications of coincident stereo or M-S recording, a stereo microphone may be the best choice. A stereo microphone normally embodies two closely spaced capsules arrayed one over the other, with individual pattern adjustment for both capsules and provision for rotating one of the capsules with respect to the other. This degree of electrical and mechanical flexibility is necessary in order to cover all aspects and conditions of coincident recording.

Departing slightly from the notion of capsule coincidence, a number of dual-microphone mounting techniques have been developed for stereo recording and are embodied in acoustically appropriate structures. Most of these are spherical, some purposely head-shaped, while others are merely some kind of baffle with microphones on opposite sides. While most of these embodiments work well enough for direct stereo pickup, some may be more suited to the requirements of binaural recording.

In addition, there are a handful of compound microphones that have been developed to pick up sound via four or more capsules for purposes of surround sound pickup; these will be discussed in a later chapter. In this chapter we will discuss the development of the conventional stereo microphone and describe several commercial embodiments in detail.

BLUMLEIN'S EARLY EXPERIMENTS

Figure 12–1 shows how Blumlein (1931), using only a pair of omnidirectional microphones, synthesized a pair of cardioid patterns splayed at ±90° below about 700 Hz, working in conjunction with a pair of left-right omnis shadowed by an absorptive baffle that separated them. The shaded omnis were effective only at frequencies above about 1500 Hz. This was in fact the first coincident stereo microphone array, and very likely the first instance of generating a cardioid pickup pattern via the electrical addition of figure-8 and omnidirectional components.

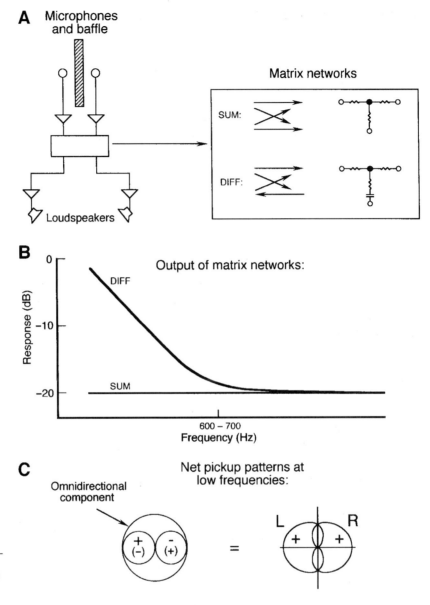

FIGURE 12-1

Blumlein's stereo recording
technique of 1931.

ANATOMY OF A STEREO MICROPHONE

The first commercial stereo microphones appeared in the fifties and were
made by Georg Neumann and Schoeps in Germany and AKG Acoustics in
Austria. Two well-known current models are shown in Figure 12–2. The
cutaway detail of the Neumann SM-69 shows the vertical arrangement of
the Braunmühl-Weber dual diaphragm capsule assemblies, with the upper
one free to rotate as required. The AKG Acoustics model C-426 is
equipped with a set of lights to aid in azimuth adjustments in field setup.

FIGURE 12-2 ———

Stereo microphones based
on the Braunmühl-Weber
dual diaphragm assembly;
the Neumann model
SM-69 (A); the AKG
Acoustics model 426 (B).
(Photos courtesy of
Neumann/USA and AKG
Acoustics.)

Figure 12–3 shows the Royer model SF-12 ribbon microphone. The model consists of a pair of ribbon elements permanently mounted at 90° and dedicated to the Blumlein crossed figure-8 format. It has an XLR five-pin output receptacle that fans out to a pair of XLR three-pin receptacles for normal console use.

The Josephson C700S is an unusual model that contains an omni element and two figure-8 elements fixed at angles of zero and 90°. Because of this, it can perform as a stereo microphone or as a first-order surround microphone. The three capsule outputs are combined in a flexible control unit to provide five simultaneous signal outputs. The choices here are all first-order patterns individually ranging from omnidirectional to figure-8. The microphone is shown in Figure 12–4A, and the signal flow diagram of the controller is shown at B.

Operating in a stereo mode, the system provides an X-Y pair of outputs whose patterns can be continuously varied from omni to figure-8 and whose included angle can vary from zero to 180°, with all adjustments made entirely at the control unit. Operating in an M-S mode, a

FIGURE 12-3 ————

Royer model SF-12 ribbon microphone. (Photo courtesy of Royer Laboratories.)

forward-oriented pattern of the user's choice and a side oriented figure-eight can be derived.

Operating in a surround mode, or as an adjunct to a larger studio pickup plan, up to five outputs can be derived. An interesting application here would be arraying the five outputs at angles of 72° and setting all patterns roughly to hypercardioid. This could be used to achieve horizontal (azimuthal) plane pickup of ambient sound.

In this regard, the C700S behaves virtually like a Soundfield microphone operating in the horizontal plane. (See Chapter 15 under The Soundfield Microphone.)

REMOTE CONTROL OF STEREO MICROPHONES

Stereo microphones that use dual diaphragm elements are normally operated with a remote control electronics unit that provides phantom powering and switchable pattern control. In most designs, the microphone capsule assemblies themselves are rotated manually. Many remote control units provide for Left/Right outputs or Mid/Side outputs, at the choice of the user. Figure 12–5A shows front and rear photos of the Neumann CU 48i remote control unit, with a circuit diagram shown at B.

As a matter of convenient interface, there are two outputs from the control unit terminating in conventional XLR-M connectors that accommodate P-48 phantom powering from the downstream console. The microphone assembly itself is fed by a shielded 10-conductor cable which is available in several interconnecting lengths to adapt to a variety of installation requirements.

HEADS, SPHERES, AND BAFFLES

At one time, an accurately scaled artificial head containing two pressure microphones at ear position was thought of only in the context of binaural recording. The basic binaural technique is shown in plan view in Figure 12–6A, and a typical binaural head is shown at B. Here, the three-dimensional sound field in the original recording space is accurately mapped into the listener's auditory domain via correct time cues and frequency response shaping around the head, along with equalization to compensate for the secondary signal path between headphones and the ear drums. Artificial head assemblies are manufactured by, among others, Georg Neumann GmbH, Knowles Electronics, Brüel & Kjaer, and Head Acoustics GmbH. In addition to binaural recording, artificial head systems are widely used in psychological acoustics and evaluation of performance spaces.

Binaural recordings are intended for headphone reproduction and do not normally translate well in direct stereo playback, primarily because of the limited channel separation at low frequencies when auditioned over stereo loudspeakers. It is possible, however, to convert a

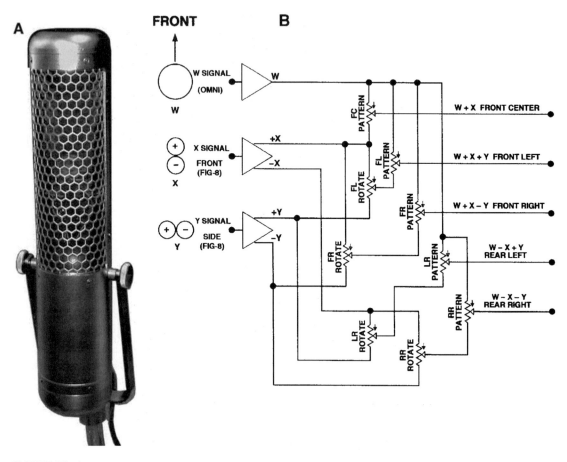

FIGURE 12-4

Details of the Josephson Engineering model C700S stereo/surround microphone system;
photo of microphone (A); signal flow diagram (B). (Data courtesy of Josephson Engineering.)

binaural recording into stereo by using a crosstalk canceling method,
such as that shown in Figure 12–6C. Because the head is small with
respect to the wavelengths of frequencies below about 700 Hz, sound
bends around the structure with little attenuation. There is however a
significant phase difference at the two microphones due to their spacing,
and this phase difference can be effectively converted to an amplitude
difference by applying equalization along with sum and difference networks. The technique has much in common with the basic Blumlein
setup shown in Figure 12–1.

Some manufacturers offer spherical baffles. These are similar to artificial heads, but are normally used directly for stereo recording with no
requisite LF crosstalk cancellation. Significant work in the development
of these recording methods has been carried out by Günther Theile
(1991). A further distinction is that their pressure microphones are

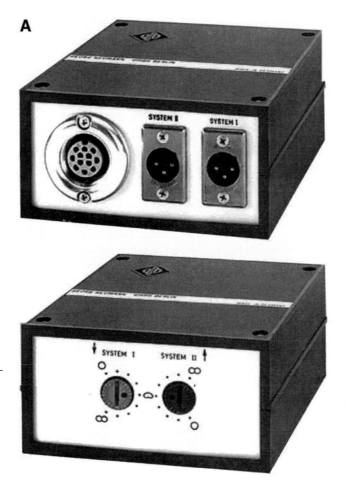

FIGURE 12-5

Stereo microphone remote control unit: photo of Neumann CU 48i remote pattern selector (A); signal flow diagram of CU 48i (B). (Data courtesy of Neumann/USA.)

mounted directly on the spherical surface, resulting in relatively flat response. A typical example of this is the Neumann KFM-100 system shown in Figure 12–7A. Frequency response and polar response in the horizontal plane are shown in Figure 12–7B and C.

Figure 12–8A shows a view of the Crown International Stereo Ambient Sampling System™ (SASS). Small omnidirectional microphones are positioned on each side at the apex of the foam-lined structure, roughly at ear distance. The shading of the unusual structure results in the polar patterns shown at B, which indicate a gradual transition above 500 Hz to a clear ±45° pair of pickup patterns above about 3 or 4 kHz.

Various forms of flat baffles have been used over the years to increase separation between near-coincident stereo microphones at high frequencies. In the modern stereo era, Madsen (1957) appears to have been the first to make use of such a baffle, as shown in Figure 12–9A. A modern example here is the Jecklin disc (1981), shown at B. Here, an acoustically damped disc with a diameter of 250 mm (10 in) is placed between two omnidirectional microphones. For sounds originating at

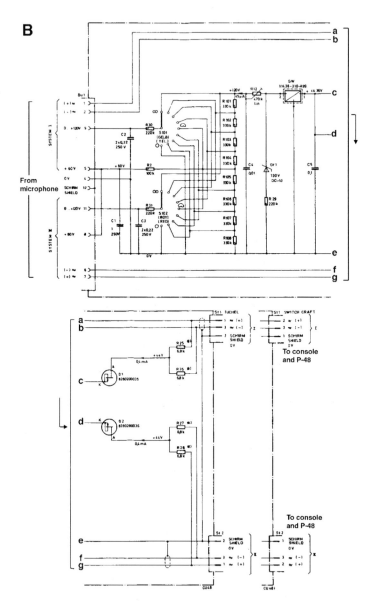

FIGURE 12-5

Continued.

90° to the array the delay between the two microphones will be about 0.7 ms, which is only slightly greater than the equivalent delay around the human head. Shadowing effects for the far-side microphone will be somewhat more complicated than for a head or a sphere, due to the rather complex nature of diffraction effects around a plane as opposed to a spherical or other smooth three-dimensional surface.

There are many competing stereo systems embodying spheres and many types of baffles. The differences among them can be fairly subtle and have to do basically with dimensions (most do not depart from

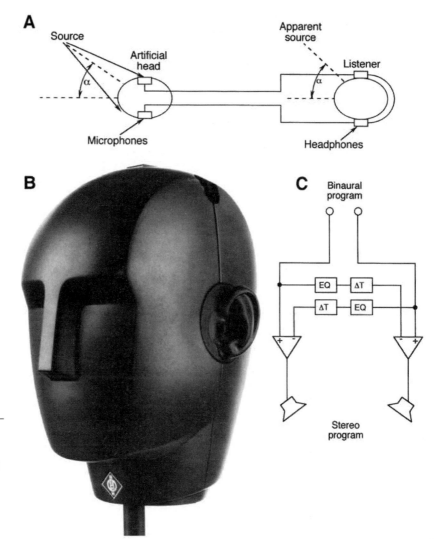

FIGURE 12–6

Basic binaural recording
signal flow diagram (A);
photo of the Neumann KU
100 binaural head (B);
binaural to stereo
conversion (C). (Photo
courtesy of
Neumann/USA.)

normal head dimensions significantly), materials (absorptive or not), and
careful equalization in the microphone channels.

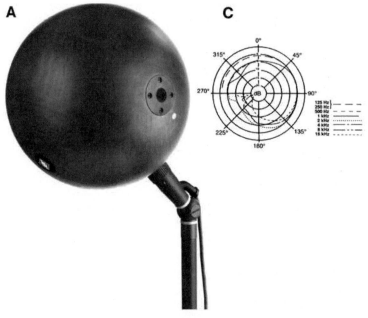

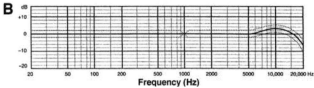

FIGURE 12-7

Neumann KFM 100 sphere
(A); frequency response (B);
polar response (C). (Data
courtesy of Neumann/USA.)

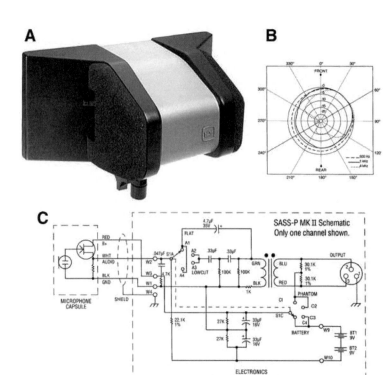

FIGURE 12-8

The Crown Internal SASS
system: photo of structure
(A); typical polar response
(B); circuit diagram of
system (C). (Data courtesy
of Crown International.)

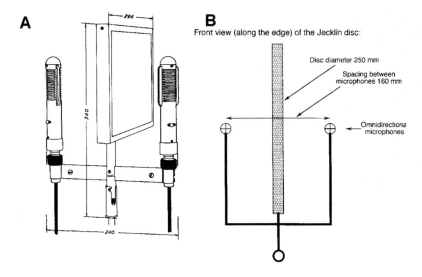

A

B

Front view (along the edge) of the Jecklin disc:

Disc diameter 250 mm

Spacing between microphones 160 mm

Omnidirectional microphones

FIGURE 12-9

Use of baffles in stereo microphone pickup: Madsen's application with a pair of ribbon microphones (A); details of the Jecklin/OSS disc (B). (Figure at A courtesy of Journal of the Audio Engineering Society.)

C H A P T E R   1 3

CLASSICAL STEREO RECORDING TECHNIQUES AND PRACTICE

INTRODUCTION

The topics covered in the previous two chapters form the basis for most of our study in this chapter. We begin with a basic discussion of the physical characteristics of musical instruments, moving on to a discussion of case studies with reference recordings made by the author. We progress from solo instruments, through chamber ensembles, to large orchestral resources, with emphasis on the practical decisions that every recording project requires. A final section deals with the adjustment of room acoustical conditions to fit the recording project at hand.

Our recommendations for microphone placement throughout this chapter remain the same whether you are recording direct-to-stereo or to multitrack. The difference of course is that multitrack offers unlimited opportunities to fix your mistakes later – rather than fixing them with retakes during the sessions.

CLASSICAL RECORDING: ART FORM OR SCIENCE?

One of the first things an aspiring recording engineer learns is that the best seats in a concert hall are not necessarily the best places to put microphones. Those seats after all are where, by common agreement, orchestral balance is judged to be ideal. Many of us have tried at least once to place microphones at favored listening positions and found that the recording sounded too reverberant, was dull at high frequencies, had insufficient stereo stage width and, during quiet passages, was somewhat noisy.

Why should this be so? As we listen in a concert hall we are able to distinguish direct sound cues (those that first reach our ears from the stage), early reflections (those that help define the apparent stage width), and global reverberant cues (those that give an impression of the size of the performance space). It is also an advantage that we can turn our heads from side to side in order to further reinforce these cues and ultimately to "zero in" on stage events and hear them in a familiar environment, even in the presence of a reflected sound field that may be several times greater in terms of acoustical power than the direct sound field itself.

A pair of stereo channels will have a difficult time duplicating this, since so wide a range of desirable spatial cues cannot be reproduced as such with only two loudspeakers. There are also performance constraints in the consumer playback environment, be it one's living room, automobile, or portable Walkman.

As a result, modern classical recording practice remains what it has always been: a close-up event with the principal microphones placed at Row A and a playback effect in the home that seems to give the impression of being seated at Row J. How this is done will become clear as this chapter progresses. First, we need to understand some of the acoustical characteristics of musical instruments, including their directional aspects and dynamic ranges.

SOME ACOUSTICAL PROPERTIES OF MUSICAL INSTRUMENTS AND ENSEMBLES

In general, instruments radiate forward and upward relative to the player's seating position; however, all musicians appreciate the reflectivity of an uncarpeted performing area. At low frequencies all instruments are essentially nondirectional. As we progress upward in frequency, radiation becomes more pronounced along the bell axes of brass instruments, while the directionality of woodwind instruments becomes fairly complex. Figures 13–1 and 13–2 illustrate this.

String instruments radiate in a complex manner. At low frequencies they are nearly omnidirectional, and at mid-frequencies there is a preference for radiation perpendicular to the top plate, or "belly", of the instrument. At the highest frequencies the radiation pattern becomes broad again as radiation is largely from the very small bridge of the instrument. These general trends are shown in Figure 13–3.

The various keyboard and percussion instruments generally have complex radiation characteristics due to their shapes and sizes and often interact significantly with their environments.

Figure 13–4 shows the dynamic ranges characteristic of the major instrumental groups. Within a given frequency range, the dynamic capability may be no more than 35 or 40 dB at most; however, over a large frequency range the total dynamic capability can be considerable. The French horn is probably most notable here, covering a range of about 65 dB from the lowest to highest frequency ranges, but limited to about 35 dB within any given frequency range. Of all orchestral instruments the clarinet has the widest overall dynamic range within a given frequency range, nearly 50 dB in the instrument's middle range.

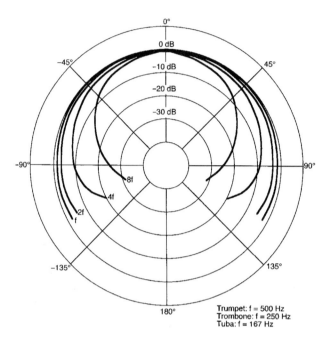

Trumpet: f = 500 Hz
Trombone: f = 250 Hz
Tuba: f = 167 Hz

FIGURE 13-1

Polar response of brass
instruments along the bell
axis; values of f are given
for three brass instruments.
(Data after Benade, 1985.)

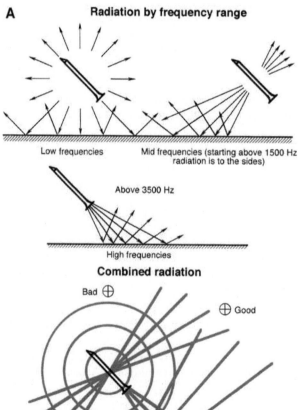

FIGURE 13-2

Directionality of
woodwind instruments:
basic characteristic (A);
off-axis measurement
graphs (B and C). (Data
after Benade, 1985.)

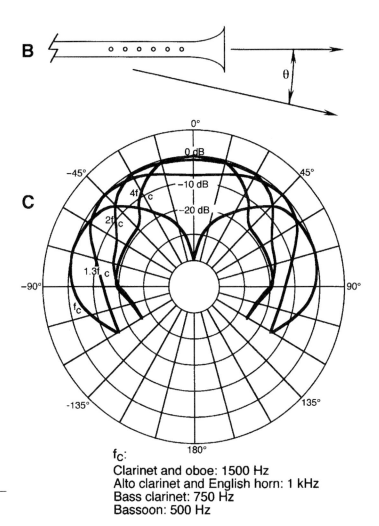

FIGURE 13-2

Continued.

f_C:
Clarinet and oboe: 1500 Hz
Alto clarinet and English horn: 1 kHz
Bass clarinet: 750 Hz
Bassoon: 500 Hz

The following discussions will largely detail the author's experiences in recording more than 280 compact discs. These are not the only approaches, and the aspiring engineer is encouraged to experiment with other approaches. References made to coincident pickup apply as well to near-coincident pickup and to head-related techniques, as discussed in preceding chapters.

RECORDING SOLO INSTRUMENTS

KEYBOARD INSTRUMENTS

One of the biggest problems in recording the piano may be the condition of the instrument itself. Most good concert grand pianos are associated with performance venues or schools of music and as such may be voiced

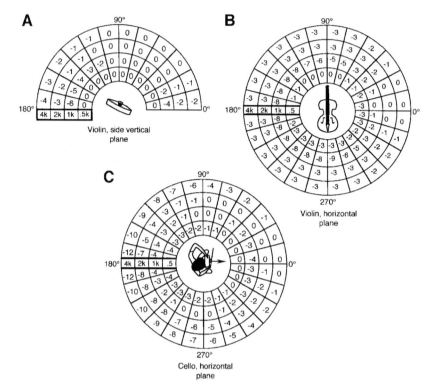

FIGURE 13-3

Directionality of string instruments: violin in transverse vertical plane (A); violin in azimuthal plane (B); cello in horizontal plane (C).

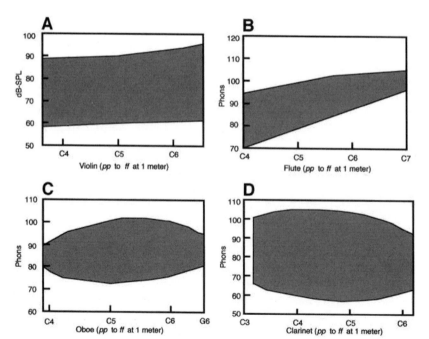

FIGURE 13-4

Dynamic ranges versus frequency; violin (A); flute (B); oboe (C); clarinet (D); trumpet (E); French horn (F).

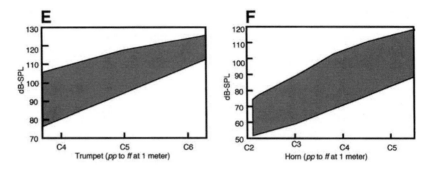

FIGURE 13-4

Continued.

for projection in recital or concert performance spaces. Often, this means that the piano may sound a little aggressive close in. If this is the case, then a technician should be asked to tone down the instrument to whatever degree may be necessary. Normally, this requires that only the upper three octaves may need to be adjusted.

The piano can be recorded best in recital halls of moderate reverberation time, as opposed to a typical concert hall. Some microphone approaches are shown in Figure 13–5; A and B show coincident pickup while C and D show spaced omnidirectional microphones. With spaced microphones, it is important not to overdo the spacing, and you should take care that the microphones be approximately equidistant from the sounding board to ensure reasonably coherent pickup of the lower vibration modes of the sounding board. Listen to various spacings before you commit yourself.

In most cases it would be a mistake to move much farther from the instrument than shown in the figure, the danger being pickup of too much reverberation. When using a coincident approach, take the time to experiment with various microphone splay angles and pickup patterns with the aim of achieving the proper stereo stage width.

Some producers and engineers prefer to remove the piano's cover and place the microphones nearly overhead. This will result in increased low frequency response and may or may not be appropriate for the repertoire. Proceed with caution.

If possible, a stand-by technician-tuner should be on hand during the sessions to do touch-up tuning or take care of any mechanical noises that may develop in the instrument. Damper pedals are often noisy and require attention. Often, a small piece of carpet under the pedal lyre will "tame" the noise produced by an overactive peddler.

The harpsichord creates a shower of higher harmonics and relatively little fundamental. As such, it can sound with great clarity in the fairly reverberant environments which are normally chosen to enhance baroque and classical literature for the instrument. The same general pickup approach for the piano may be used here, with a slight preference for spaced omni microphones. The harpsichord key action may generate

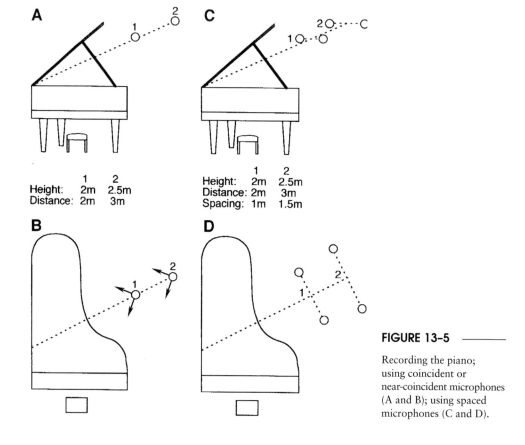

FIGURE 13-5

Recording the piano;
using coincident or
near-coincident microphones
(A and B); using spaced
microphones (C and D).

a good bit of noise, which can usually be alleviated with a 60 or 80 Hz
high pass filter.

Regarding the stereo soundstage presentation, a piano should appear
to be spread over a center portion about two-thirds the total stage width,
with reverberant cues coming from the entire stage width. (See Compact
Disc Reference 1.)

THE GUITAR AND LUTE

Again, the choice is between coincident and spaced microphones, and the
basic options are shown in Figure 13–6. With coincident pickup, watch
out for proximity effect, and be prepared to equalize it accordingly. If the
primary microphones do not pick up enough reverberation, you can use
a spaced pair of room microphones to enhance the sound. Do this only
if the room is quiet enough; otherwise, consider using a high quality digital reverberation generator with appropriate room parameters pro-
grammed into it. (See Compact Disc Reference 2.)

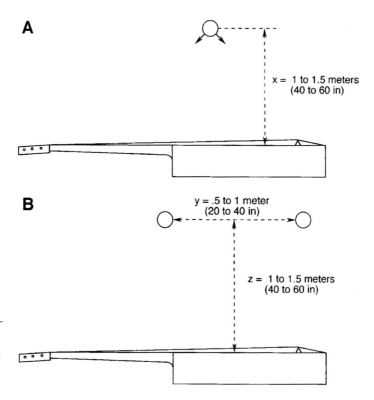

A

x = 1 to 1.5 meters
(40 to 60 in)

B

y = .5 to 1 meter
(20 to 40 in)

z = 1 to 1.5 meters
(40 to 60 in)

FIGURE 13-6

Recording the guitar: using coincident or near-coincident microphones (A); using spaced microphones (B).

THE ORGAN

Of all instruments, the organ is the only one that does not "travel." Unless it is a small *portative* instrument which is often used for baroque music accompaniment, the organ is permanently built into a large space and voiced for that space. No two organs are even remotely alike, and the choice of venue is equally important as the choice of performer and literature. The organ is usually heard at a large distance, and specific directional effects from the instrument are the exception. The listener thus hears the organ and its environment as an entity, rather than as an instrument independent of its surroundings.

In many cases a single pair of omnidirectional microphones will suffice to pick up the instrument at a distance up to 6 or 10 m (20 to 33 ft). If the instrument's keyboard divisions are laid out horizontally, then a coincident pair can be used to delineate lateral directionality. Usually the instrument is arrayed vertically, and lateral imaging will be lost as such. In some cases an instrument installed in the rear gallery of a church consists of a main set of pipework along with a *rückpositiv* division suspended on the gallery railing. In this case the fore-aft imaging aspects will be readily apparent in the recording. A suggested starting point is shown in Figure 13–7.

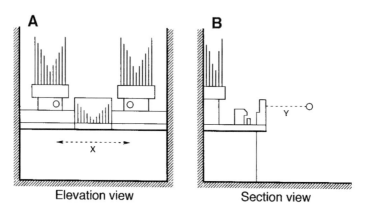

FIGURE 13-7 ————

Recording the organ:
elevation view (A); section
view (B).

In many European cathedrals the organ is located very high in the
rear gallery, as much as 15–20 m (50–65 ft) above the floor of the nave.
There are no microphone stands tall enough to put the microphones
where they need to be, and the normal approach is to hang microphones
from the ceiling or from lines suspended from side to side at the height
of the triforium gallery. Such preparations take time and can result in
microphone cable runs up to 100 or 150 m (330 to 450 ft).

If the engineer is required to work from the floor it may be useful to
experiment with both omnidirectional microphones on high stands,
along with a widely spaced pair of line microphones aimed at the instrument. These will provide added directionality at high frequencies and
may help to clarify the texture of the instrument.

A secondary pair of microphones in the middle of the venue may
flesh out the reverberant signature of the space, should that be needed.
These microphones may be rolled off at low frequencies to minimize LF
room rumble and to maintain clarity. In some spaces that are large but
not sufficiently reverberant, an artificial reverberation generator may be
useful. However, not many reverberators can duplicate accurately the
long reverberation time associated with large spaces. (See Compact Disc
Reference 3.)

RECORDING CHAMBER MUSIC

Chamber music may be roughly defined as music written for groups
numbering from two to perhaps twelve performers, with one performer
on each part and normally performing without a conductor. For smaller
chamber groups there are basically two choices for the recording engineer: to record the group in a concert setting as they would normally be
seated on-stage; or to record them in a studio setting arrayed in an arc,
or even facing each other. There are advantages to both approaches.

The concert setup will be familiar to the players and as such may be
preferred at the outset. Conventional microphone placement would call

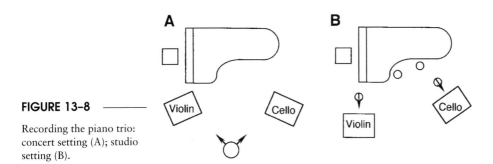

FIGURE 13-8

Recording the piano trio: concert setting (A); studio setting (B).

for a coincident pair, perhaps with an additional flanking omnidirectional pair. This approach will make the ensemble sound "staged"; that is, the listener will hear the group with a "Row G" perspective. If more intimacy is desired the group can be placed in an arc or even a circle, with microphones more closely placed. Figure 13–8 illustrates both approaches. At A we see a piano trio (piano, violin, and cello) in a normal concert setup, with a frontal array of microphones. At B we show an alternate approach of positioning the violin and cello players so that all three members of the group now have direct eye contact. (See Compact Disc Reference 4.)

A typical listener reaction to these two recording perspectives is that the former places the listener in the concert hall ("you are there"), while the latter places the performers in the listener's living room ("they are here"). The players, once they have adjusted to the alternate seating, usually become partial to it. The engineer and producer favor it because it gives them more flexibility in establishing and altering balances as the recording gets underway.

Recording a solo instrument or vocalist with piano is greatly helped by departing from the concert setup, as shown in Figure 13–9. Here, the vocalist has been placed a comfortable distance from the piano and is facing the pianist, thus ensuring good eye contact. Both piano and soloist are individually picked up in stereo, and a mix is created in the control room. If good judgments are made in balance and placement, everything will sound perfectly natural. This approach allows the engineer and producer to achieve the desired balance between soloist and piano at all times.

The question of when to use the piano with its cover on half-stick often comes up quite often. In concert, half-stick may be necessary to keep the instrument from overpowering soloists. However, in the studio environment, it is preferred to keep the cover at full-stick and achieve the necessary balances purely through placement of microphones and performers. (See Compact Disc Reference 5.)

The string quartet is one of the most enduring musical ensembles of all time, and the medium has a rich and complex repertoire that continues to grow. Normally the players are seated as shown in Figure 13–10A. While the traditional frontal coincident array is often used, as shown at B, many engineers favor an array of spaced omnidirectional microphones,

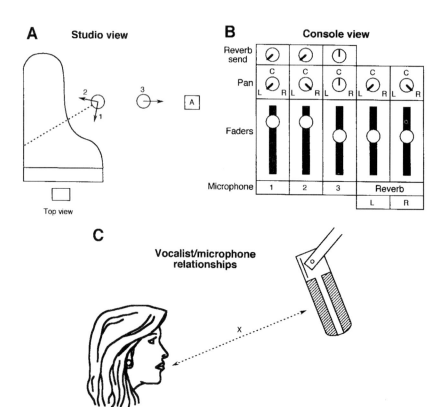

FIGURE 13-9

Recording vocalist with
piano in the studio:
microphone positions (A);
console assignments (B);
vocalist's orientation to
microphone (C); distance x
is normally in the range of
0.5 m (20 inches) to 1 m
(40 inches) (C).

as shown at C. The center microphone would normally be mixed in at a
level about -4 to -6 dB relative to the two others and panned just
slightly right of center. Its purpose is to anchor the center of the stereo
stage (primarily the cello) without appreciably narrowing the overall
stereo stage width. Many engineers use a cardioid microphone for the
center pickup. The two flanking omnis should be positioned so that they
preserve the balance of the entire group. (See Compact Disc Reference 6.)

THE CHAMBER ORCHESTRA; INTRODUCTION TO ACCENT MICROPHONES

The chamber orchestra normally numbers from 25 to 40 players,
depending on the repertoire. The majority of the literature for the chamber orchestra is from the late 18th and early 19th centuries. The ensemble is best recorded with four microphones across the front, as discussed
in the previous chapter, with the addition of *accent* microphones as
needed. Accent microphones, also familiarly called "spot mikes", are
used to add presence to an instrument or an orchestral section without
necessarily increasing loudness. On a small reflective stage they may not
be needed at all. But under most conditions we would use a secondary

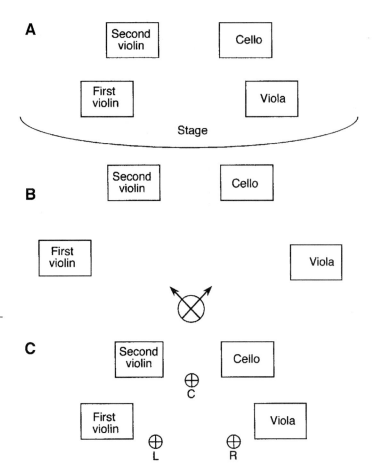

FIGURE 13-10

Recording the string quartet; normal concert seating (A); use of coincident or near-coincident microphones (B); use of spaced omnidirectional microphones with an accent microphone for the cello (C).

coincident stereo pair for picking up the woodwind section, along with single accent microphones for the first two stands of bass viols, harp and celesta. The timpani are normally loud enough, but an accent microphone may help bring out the articulation of the drums. A typical seating arrangement is shown in Figure 13–11. (See Compact Disc Reference 7.)

Accent microphones are normally cardioids and are generally mixed into the main array at levels according to the technique shown in Figure 13–12. Many engineers prefer to delay individually all of the accent microphones so that their signals will be essentially time-coherent with the pickup of the main pair, as shown at A. That is, if an accent microphone is located, say, 8 m from the main pair, it may be delayed by 8/344 seconds, or about 23 ms, to align the pickup with its acoustical signal path to the main pair. While it is always correct to delay accent microphones, it may not be necessary because of the masking effect of the louder pickup from the main microphone array. Data shown at B indicate approximately when delay is mandatory, while data shown at C show the approximate level relationships between accent microphones and main microphones.

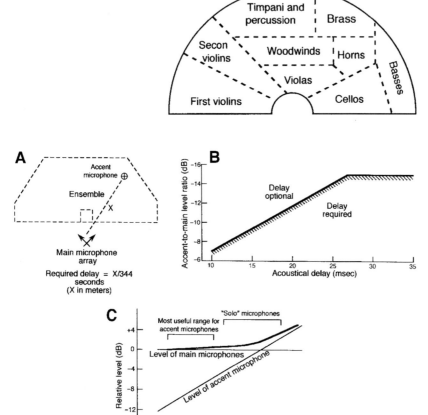

FIGURE 13-11 ————

Recording the chamber
orchestra; normal seating
arrangement.

FIGURE 13-12 ————

Implementing accent
microphones: calculating
microphone delay (A);
determining when delay is
and is not essential (B); level
range for accent
microphones (C).

Remember that it is very important to pan the positions of the accent microphones so that they correspond exactly to the physical position of the instruments as heard on the stereo stage when no accent microphones are engaged. It is also common practice to attenuate slightly the low frequency response of the accent microphones; this makes it easier to add presence without adding loudness.

THE LARGE ORCHESTRA

The full symphony orchestra may have a complement of strings of 14-12-10-8-8 (respectively, the number of first violins, second violins, violas, cellos, and basses). Woodwinds may number from 8 to 16, depending on the nature of the score. French horns may number anywhere from 4 to 12, depending on the score, and the heavy brass may number 4 trumpets, 3 trombones, and a single tuba. The percussion resources, including timpani, may call for 4 players. Add to all of this the requirements of perhaps two harpists and a keyboard player (celesta and/or piano), and the head count can easily reach or exceed 90 players.

A symphony orchestra takes up much floor space, and a good rule is to multiply the number of players by two to arrive at the approximate number of square meters of total occupied floor space, taking into account a normal complement of players in each section. Many modern stages are not quite large enough to fit this paradigm, and many orchestral recordings routinely take place instead in large, fairly reverberant ballrooms or other large meeting spaces. This is especially true in England and on the European continent. Churches are often called into duty for large-scale recording, but the excessive reverberation time of many of these venues can be a problem.

MICROPHONE PICKUP ANALYSIS

Figure 13–13 shows plan and side views of the recording setup for a live (with audience) recording of Gustave Holst's "The Planets", a typical large orchestral work dating from 1916. Note that there are 14 microphones. Of these, the four across the front provide the basic pickup; all others are accent microphones.

The main four microphones consist of an inner ORTF pair with a flanking pair of omnidirectional microphones. The ORTF pair is aimed downward at a point about two-thirds the depth of the stage, and the intent here is to avoid picking up too much of the front of the orchestra by relying on the off-axis attenuation of the cardioid patterns to balance the front-back distance ratio.

The accent microphones are detailed below:

1. Harps: two instruments with a single cardioid placed between them.
2. Celesta: a single cardioid to delineate this soft instrument.
3. Horns: a single cardioid placed about 4 m (13 ft) *above* the section and aimed downward toward the edge of the bells.
4. Woodwinds: an ORTF pair aimed at the second (back) row of players.
5. Timpani: two sets of drums were used, and a single cardioid microphone was placed midway between them about 2.5 m (8 ft) above the stage.
6. Brass: a single cardioid placed about 4 m above the players, aimed downward.
7. Basses: a single cardioid placed overhead about 1.5 m (5 ft) from the first stand of players.
8. House pair: two widely spaced cardioids hanging from the attic and positioned about 7 m (23 ft) above the audience and about 8 m (26 ft) from the edge of the stage. These microphones were aimed at the upper back corners of the hall.

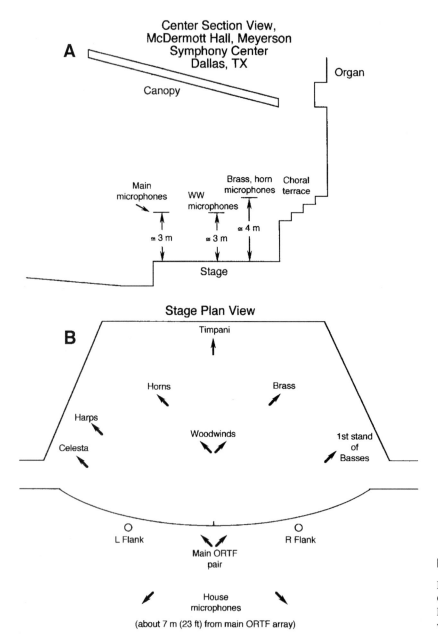

A Center Section View, McDermott Hall, Meyerson Symphony Center Dallas, TX

Canopy

Organ

Main microphones
WW microphones
Brass, horn microphones
Choral terrace

≅ 3 m   ≅ 3 m   ≅ 4 m

Stage

B Stage Plan View

Timpani

Horns   Brass

Harps

Celesta   Woodwinds   1st stand of Basses

L Flank   R Flank

Main ORTF pair

House microphones

(about 7 m (23 ft) from main ORTF array)

FIGURE 13–13 ————

Dallas Symphony Orchestra, recording Holst's *The Planets*: Side view (A); plan view (B).

All accent microphones were operated with a slight LF rolloff below about 100 Hz. With the exception of the house pair, they were all fed in varying amounts to a digital reverberation unit whose parameters were set to match those of the occupied hall. The stereo output of the reverberation unit was returned to the two-channel stereo mix at an appropriate level – no more than to ensure that the effect of the accent microphones blended with the natural pickup provided by the main four microphones across the front of the orchestra. An off-stage chorus was located in the stage house behind the orchestra and the sound was introduced into the hall via the stage doors. No additional microphones were used here.

The choice of instruments and sections to be given preference with accent microphones was made in consultation with the recording producer, taking into account overall musical requirements of the score. While the main microphones were virtually fixed in level, the individual accent microphones were adjusted on a running basis by no more than about ±2.5 dB, according to musical requirements. (See Compact Disc Reference 8.)

SOME VARIATIONS ON THE LARGE ORCHESTRA SETUP

In many modern concert halls a large chorus may be placed in a choral terrace behind the orchestra, calling for an additional stereo pair of microphones. In many older halls, the orchestra is moved forward and the chorus placed on risers positioned behind the back row of orchestra players. If the hall is of the proscenium type, the chorus often finds itself in the very dead up-stage acoustical setting, calling for its own microphones (two or three in stereo) along with the correct amount of artificial reverberation to match them with the normal orchestra pickup.

A recurring problem with recording works for chorus and orchestra is leakage of the back of the orchestra (brass and percussion) into the choral microphones. This can be alleviated by elevating the choral microphones as high as is practicable by overhead rigging.

If there is no audience present, an excellent alternative is to place the chorus in the house *behind* the conductor and picking it up with its own dedicated stereo microphone array. The effect is excellent in that the chorus has immediate acoustical contact with the hall.

COMMENT ON FRONT–BACK COVERAGE OF THE ORCHESTRA

The main and woodwind ORTF microphone pairs are aimed downward in order to adjust front–back balances in their respective pickup patterns. This procedure is detailed in Figure 13–14.

The directional patterns are of course cardioid, and the idea here is to trade off the balance of the nearer instruments with those toward the rear by taking advantage of the off-axis attenuation of the microphones' patterns. Considering the main ORTF pair, as shown in Figure 13–14, the elevation angle has been set so that the primary axis of the microphones is aimed at the woodwinds, brass and horns. When this is done, the off-axis pattern attenuation of the frontal strings will be about −2.5 dB. However, the front of the orchestra is about one-third the distance from the microphones relative to the more distant woodwinds, brass and horns. The net pickup difference will then be 9.5 − 2.5, or about 7 dB, favoring the front of the orchestra. We must remember however that both woodwinds and brass instruments are essentially forward facing with

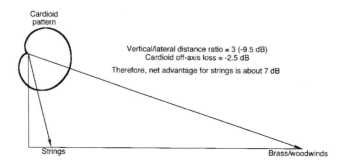

FIGURE 13–14

Downward elevation of main ORTF microphones.

directivity indices that are in the range of 4 or 5 dB. Thus, with their direct sound virtually aimed at the frontal microphones, the front–back orchestra balance will generally seem quite natural. In some cases both engineer and producer may wish to emphasize the strings even more by lowering the elevation angle to favor the strings to a greater degree.

SOLOISTS WITH ORCHESTRA

In a concert setting the addition of a piano for concerto use will require that the center-front orchestral players move upstage as needed so that the piano may be placed between the conductor and the house, as shown in Figure 13–15A. If an audience is present, this repositioning often interferes with the placement of the main microphone pair. Ideally, it should be just in front of the piano, but considerations of sightlines may force the engineer to place the microphone stand just behind the instrument. With hanging microphones there is no problem here at all; the main pair can be placed in its ideal position. (See Compact Disc Reference 9.)

Vocal or string soloists are often placed to the conductor's left and will not interfere with the main microphone setup. Whenever possible, these soloists should be picked up with a coincident or near-coincident stereo pair, as opposed to a single microphone. The intent here is to preserve natural stereo perspectives in case the engineer has to increase the level of the soloist's microphones during very soft passages. Details are shown in Figure 13–15B. (See Compact Disc Reference 10.)

LIVENING OR DEADENING THE PERFORMANCE SPACE

A hall that is too live can be easily deadened by placing velour material at strategic points in the performance space. Where possible, the velour should be hung over balcony railings or freely hung away from the walls as banners. A little will often go a long way, and it is highly recommended that an acoustical consultant be engaged to oversee this temporary modification.

It is a bit more complicated to liven a space, but the technique detailed in Figure 13–16 produces surprising results. Before and after examples are

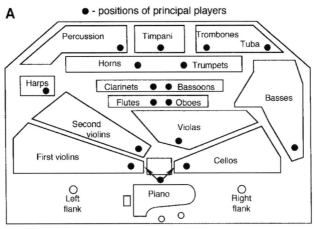

Main ORTF pair must be moved as shown to a position directly behind the piano.

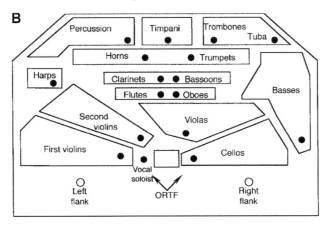

FIGURE 13-15

Recording soloists with orchestra. Piano (A); vocalist (B).

given in Compact Disc Reference 11. The effect of the 0.004 in. (0.1 mm) thick vinyl material is to reflect high frequencies arriving at a fairly low grazing angle of incidence. It does not have to be stretched in any manner; rather, simply laid over the seating area. (The vinyl material in question is available at building supply stores under a number of trade names. No material less than 0.004 in. thickness should be used.)

As we have mentioned before, artificial reverberation on accent microphones is an accepted practice in current classical recording; however, there is no substitute for natural reverberation to enhance the primary pickup via the main microphones and house microphones.

HOW MUCH REVERBERATION IS NEEDED IN THE RECORDING?

Music of different periods will require varying amounts of reverberation, both in level and in decay time. General recommendations for concert venues, as a function of room volume and music type, are given in

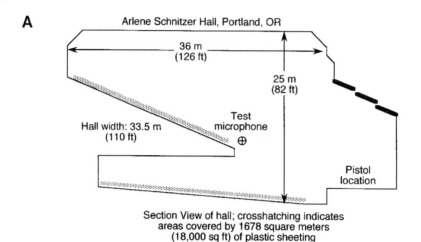

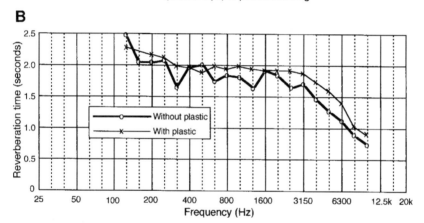

FIGURE 13-16

Livening the recording venue: section view (A); measured reverberation time with and without plastic material (B).

Figure 13–17. In general, reverberation times exceeding about 2.5 seconds will sound rather unnatural on orchestral musical, whatever the period. Both classical and modern era music fare best with reverberation times on the order of 1.5 s, while romantic compositions may do with reverberation times up to 2 s. While ecclesiastical music is often heard in large spaces with up to 4 or 5 s of reverberation time, in recording it is probably better to err on the short side in order to keep the reverberation from interfering with musical details.

Kuhl (1954) developed the data shown in Figure 13–18. Here, monophonic recordings were played for a listening panel whose individual preferences for reverberation time are plotted. The median points are taken as target values. We need to add that these tests, if repeated today in stereo, could result in slightly longer target reverberation time values, due to the ability of stereo listening to delineate direct-to-reverberant details more accurately than in mono listening. In any event, the tendencies shown here are certainly relevant in today's recording activities.

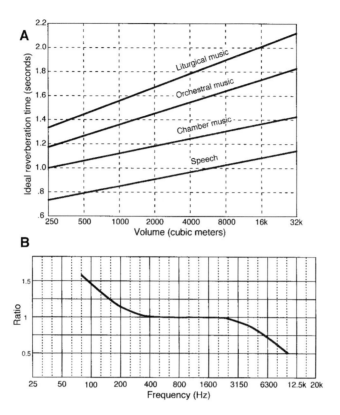

FIGURE 13-17

Typical reverberation times versus room volume for various types of music (A); normal variation of LF and HF reverberation times compared with the midband (B).

Once the target reverberation time has been established, the *amount* of reverberation introduced into the recording is critical. The sound picked up by the house microphones and introduced into the stereo mix will determine this ratio. This is an area where accurate loudspeaker monitoring conditions and experience on the part of both engineer and producer are of primary importance. In many cases, this important judgement is not made until the postproduction stage.

MAINTAINING A CONSISTENT FORE-AFT PERSPECTIVE

It is up to the engineer and producer to determine the effective "distance" of the pickup, relative to the stage, and make it convincing. Rather than placing microphones at Row J, we make the recording fairly close-in, and then we introduce reverberation to "zoom out" to where we want it to be. A slight increase in the amount of signal from the house microphones will result in a surprising increase in effective distance, so care should be taken in making these balances. When musical reasons justify this, it may be desirable to pan in the main flanking microphones very slightly toward the center, just to give the direct stage pickup a slightly narrower image to match the increase in apparent fore–aft distance. If this is not done, we run the risk of making a recording with conflicting spatial cues; it may sound both close-in and distant at the same time.

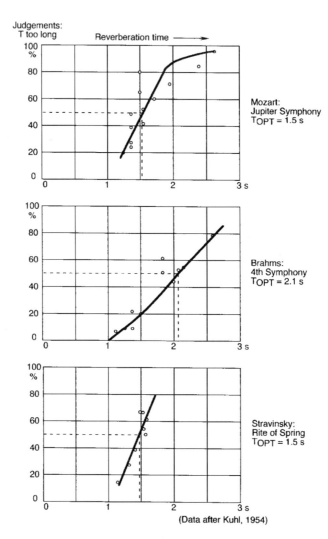

FIGURE 13-18

Kuhl's data on preferred recorded reverberation times for three orchestral works.

(Data after Kuhl, 1954)

DYNAMIC RANGE ISSUES

As we have seen in previous chapters, today's quality capacitor microphones may have dynamic ranges well in excess of 125 dB. However, in classical recording, we are rarely able to take advantage of more than about 90 or 95 of those precious decibels, as illustrated in Figure 13–19.

As we can see, the noise floor of a modern quiet venue is just about at the same effective level as the self-noise floor of the primary microphones themselves. As the amplified signals from the microphones are transmitted downstream to a digital recorder, the noise floor limitation in the recording remains that of the microphones and the room.

Thus, the choices of both venue and microphones remain all important. Microphones with self-noise in the range of 7 to 13 dB(A) are strongly recommended for all classical orchestral recording.

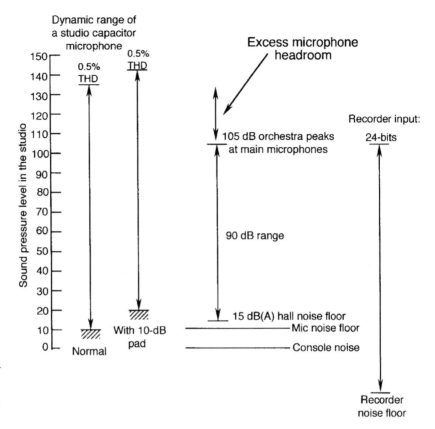

FIGURE 13-19

Dynamic range of recording microphones in a concert venue.

COMPACT DISC REFERENCES

1. *Singing on the Water*, Delos CD DE 3172, Carol Rosenberger, piano. Recorded with two spaced omnidirectional microphones.

2. *Bach: Violin Sonatas and Partitas, Arranged for Guitar*, Delos CD DE 3232, Paul Galbraith, guitar. Recorded with two spaced omni-directional microphones.

3. *Things Visible and Invisible*, Delos CD DE 3147, Catharine Crozier, organ. Recorded with two omnidirectional microphones spaced about 4 m at a distance of 10 m.

4. *Arensky/Tchaikowsky Piano Trios*, Delos CD DE 3056, Golabek, Solow, Cardenes Trio. Recorded as detailed in Figure 14–8B.

5. *Love Songs*, Delos CD DE 3029, Arleen Augér, soprano. Recording made as shown in Figure 14–9. Some room reverberation added.

6. *Mendelssohn/Grieg String Quartets*, Delos CD DE 3153, Shanghai Quartet. Recorded as shown in Figure 14–10C.

7. Haydn, *Symphonies 51 and 100*, Delos CD DE 3064, Gerard Schwarz conducting the Scottish Chamber Orchestra. Recorded as shown in Figure 14–12.

8. Holst, *The Planets* (coupled with Strauss: *Also Sprach Zarathustra*), Delos CD DE 3225, Andrew Litton conducting the Dallas Symphony Orchestra. Recorded as detailed in Figure 14–13.

9. Shostakovich, *Piano Concerto no. 2*, Delos CD DE 3246, Andrew Litton, pianist and conductor, Dallas Symphony Orchestra. Recorded as shown in Figure 14–15A, but with main ORTF piano between piano and audience.

10. Mahler, *Symphony no. 2*, Delos CD DE 3237, Andrew Litton, Dallas Symphony Orchestra and Chorus. Recorded as shown in Figure 14–15B, with two vocal soloists on either side of the conductor and with chorus located in choral terrace behind and at sides of the orchestra.

11. *Second Stage*, Delos CD DE 3504, various orchestral movements to demonstrate recording perspectives. Band 2 recorded in Portland, OR, concert hall without livening; band 12 recorded in same hall with livening technique as discussed in Figure 13–16.

C H A P T E R   1 4

STUDIO RECORDING TECHNIQUES

INTRODUCTION

Most recordings of commercial music originate not in traditional performance venues but in professional recording studios. Most studios used for pop and rock recording are large enough to accommodate no more than 12 to 20 musicians comfortably, and only in large metropolitan centers will we find large sound stage environments that can accommodate a 50- to 75-piece orchestra. For the most part, microphone placement in these studios is much closer to the instruments than is usual in classical recording, and there is much more usage of accent microphones. In virtually all cases there will be extensive use of multitrack recording, which allows for greater flexibility in postproduction.

In this chapter we will discuss a wide range of recording activities and supporting functions that routinely take place in the studio. We will also discuss aspects of studio acoustical treatment and arrangements for instrumental isolation, both individually and in groups.

We will also develop in this chapter the notion of recording as an art in its own right and not as a simple documentation of an acoustical event. Many of our recommendations in balance and panning may seem to be arbitrary, but their effectiveness in two-channel stereo playback justifies them.

The modern studio should have a modern control room, fitted with monitor loudspeakers that are symmetrically arrayed around the primary workspace. They may be built in, or they may be placed on stands. It is important that they exhibit essentially flat response over the frequency range out to about 6 or 8 kHz, with no more than a slight rolloff above that frequency. The response in both channels should be within about 2 dB of each other from about 100 Hz to about 8 kHz, and the systems should be able, on a per-channel basis, to reproduce cleanly midband signals of 105 dB at the engineer's operating position.

RECORDING THE DRUM SET

The drum set is a vital element in virtually all popular and rock music groups. While its actual makeup may vary from one player to another, the layout shown in Figure 14–1 is typical. Its basic elements are:

1. *Kick drum:* played with a spring-loaded beating attachment operated by the right foot, striking the head of the drum (the head is the stretched playing surface).

2. *Snare drum:* played by sticks or light metal brushes with both hands. The snares (tight gut strings across the rear head) are normally engaged and give the struck instrument its characteristic "snap".

3. *Hi-hat cymbals:* consists of one fixed cymbal and a movable one which is articulated by the player's left foot. The pair of cymbals are struck with a single stick in the right hand, and the left hand is used to position them, varying the timbre.

4. *Overhead cymbals:* there are usually three of these, freely struck by single sticks. The cymbals are often known by the following names: "ride" cymbal, "crash" cymbal, and "sizzle" cymbal. The last-named one has multiple small pins located loosely in holes around the periphery of the cymbal. These vibrate freely, metal-against-metal when the cymbal is struck, producing a sound rich in high overtones.

5. *Tom-toms:* Commonly known as "toms", these are small drums struck with sticks.

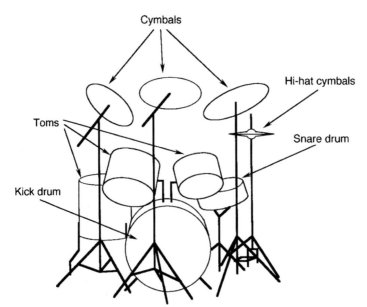

FIGURE 14-1

Front view of the modern drum set.

Anyone who has ever witnessed a studio recording will recall the drummer's long setup time and the many adjustments made to the drum set before playing gets under way. Spurious resonances and ringing must be damped out, and the player will make absolutely sure that the various moving mechanisms are silent in their operation.

Every instrument in the drum set will be precisely positioned, and the microphones, no matter how closely you position them, must not interfere with the player's movements in any way.

A SIMPLE RECORDING APPROACH

Many small acoustical jazz groups require only basic pickup of the drum set, consisting of an overhead stereo pair and a single kick drum microphone. The overhead pair will normally be cardioid capacitors with flat, extended response. Place these microphones as shown in Figure 14–2, with their axes aimed at the hi-hat cymbals and right tom, respectively. Both microphones should be placed slightly higher than the player's head

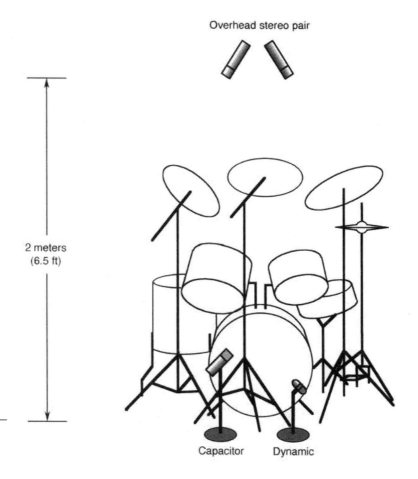

Overhead stereo pair

2 meters
(6.5 ft)

Capacitor   Dynamic

FIGURE 14–2

Simple pickup of the drum set.

and slightly to the rear so that they will not distract the player. The general aim here is to pick up the sound of the drum set very much as it is heard by the player. Some engineers prefer to use a near-coincident spacing of the overhead microphones; the wider spacing shown here will give a slightly broader stereo presentation and is recommended for that reason.

The kick drum microphone is normally placed close to the instrument on the far side from the player. Microphone choice varies; many engineers use a dynamic microphone with known low frequency headroom reserve, while others prefer a capacitor microphone. The differences here are purely a matter of taste, but in either event make sure that the microphone can handle the loudest signals the player will produce.

The drum tracks are normally recorded dry (without any signal processing in the chain), with each microphone on a separate track. In stereo presentation, the kick drum is normally panned into the center stereo position, while the overheads are panned left and right, as determined by musical requirements and spatial considerations.

Watch recorded levels carefully; the acoustical levels generated close to the drum set can easily reach the 135 dB L_P range during loud playing, so make sure that your microphones can handle these levels comfortably.

A MORE COMPLEX RECORDING APPROACH

Where recorded track capability is not a problem and where the music demands it, you can pick up individual elements within the drum set as shown in Figure 14–3. The following rules should be observed:

1. *Toms and the snare drum:* You can use small electret capacitor clip-on mikes, as shown in Figure 14–4, or you can place small diameter capacitor microphones slightly above and behind the drum. Whatever your choice, make sure that the microphone is away from any position or path that the drumstick is likely to pass through or near.

2. *Cymbals and the hi-hat:* These instruments will swing about their central pivot points when struck, and microphones must be placed carefully within the drum set away from the player and with enough clearance so that they will not interfere with playing. The normal position of the microphone should not be in line with the edge of the cymbal but rather slightly above the cymbal. The reason for this recommendation is that the radiation pattern of the cymbal changes rapidly along its edge with small angular changes. Details here are shown in Figure 14–5.

When so many microphones are used on a single set of instruments it is important to determine their real value in creating the final musical product. Needless to say, all of the microphones should be of cardioid or hypercardioid pattern in order to keep unnecessary leakage to a minimum.

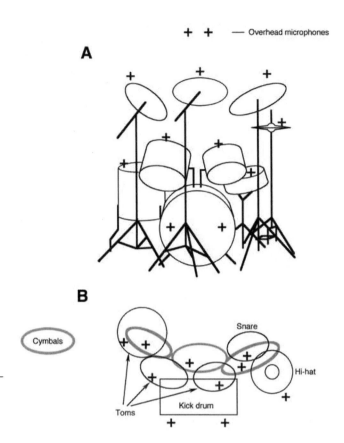

+ + — Overhead microphones

A

B

Cymbals

Snare

Hi-hat

Toms

Kick drum

FIGURE 14-3

More detailed pickup of the drum set. Front view (A); top view (B).

FIGURE 14-4

Detail of a clip-on microphone for drum pickup. (Photo courtesy of AKG Acoustics.)

If recorded channel capacity is an issue, there is always the possibility that some degree of stereo sub-grouping of multiple tracks will save on the total channel count. The engineer and producer should always think ahead and have in mind, as early as the initial tracking sessions, what the downstream postproduction requirements might be.

SOME ACOUSTICAL CONSIDERATIONS

Drums are very efficient instruments; the player does not have to expend much physical power in order to produce very high sound levels in the studio. Almost routinely, the drum kit and player are isolated from the other musicians to reduce leakage from the drum set into other open microphones in the studio. Details of a drum cage are shown in Figure 14–6. An effective cage can reduce the mid-band ambient level from the drum set by up to 15 dB in the studio. It is important not to place the drummer in an isolation booth *per se*. Eye contact is essential, as is direct speaking contact with other players. In spite of all the steps necessary to minimize drum leakage, it is important that the drummer feel a part of the playing group.

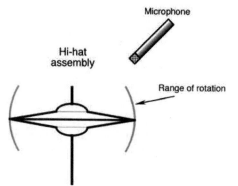

FIGURE 14-5 ————

Microphone placement
for cymbals.

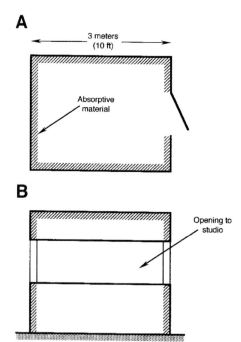

FIGURE 14-6 ————

Details of a drum
cage: plan view (A); front
view (B).

Further use of baffles (also known as "goboes") in the studio can
provide more acoustical separation among players as needed. The effectiveness of a single large baffle placed in a fairly well damped studio is
shown in Figure 14–7. Here, you can see that at mid and high frequencies the isolation is in the range of 15 to 22 dB.

During tracking sessions, it is customary to provide all players with
headphones so that "everyone can hear everyone else." Generally, each
player can have a separate headphone mix, and in some large studios an
assistant engineer, using a monitor console located in the studio, provides
monitor mixes for all players.

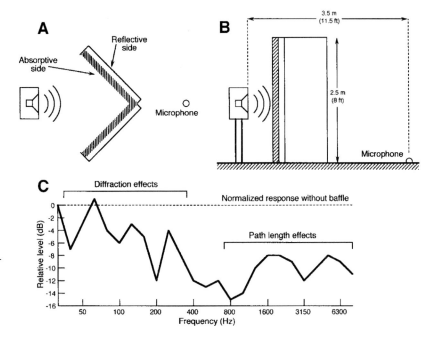

FIGURE 14-7

Effectiveness of baffles:
top view (A); side view
(B); isolation versus
frequency (C).

RECORDING OTHER PERCUSSION INSTRUMENTS

There are more percussion instruments than we can hope to detail here; however, they all fall into several basic groups, according to the principle of sound generation:

The non-pitched percussion instruments include:

- Metallophones (metal resonators):
  Triangle (struck with a small metallic beater)
  Gongs and tam-tams (struck with soft beaters)
  Finger cymbals (struck together)
  Bell tree (graduated set of nested bells, usually struck in succession with metal beater)
  Cocolo (small metal chains loosely positioned around a metal core that is rotated by the player)

- Xylophones (non-metallic resonators):
  Maracas (dried gourds filled with seeds and shaken)
  Castanets (small wood-against-wood pieces shaken together)
  Claves (hard wood sticks struck together)
  Guiro (dried gourd with serrations; played with scraper)

- Membranophones (instruments with stretched diaphragms):
  Bongo drums (set of Latin drums roughly tuned high to low)

Tambourine (combination of metallophone and membranophone)
Other ethnic drums

The pitched percussion instruments include:

- Metallophones:
    Orchestra bells (set of metal bars in keyboard layout)
    Chimes (suspended tubular bells in keyboard layout)
    Celesta (tuned metal bars played with keyboard)
    Vibraphone (tuned metal bars played with mallets)
- Xylophones (generic):
    Xylophone (short tuned wooden bars in keyboard layout)
    Marimba (tuned wooden bars with resonators in keyboard layout)

The smaller percussion instruments are normally played by two or three musicians, who move freely from instrument to instrument as called for in the musical score. It is best to record them in stereo with good left-to-right spread, assigning them to one or more stereo pairs of recorded tracks. For example, a Latin work with might call for a significant marimba part to be recorded in stereo, while another stereo pair of tracks will be assigned to a group of non-tuned percussion instruments. Normally the microphone would be about 0.5–1 m (20–40 in) away from the instruments. Recording the vibraphone in stereo is shown in Figure 14–8. The same setup applies to the marimba and other tuned mallet instruments.

Sustaining pedal

FIGURE 14-8

A widely splayed pair of cardioids for recording the vibraphone in stereo.

RECORDING THE PIANO

The piano is invariably picked up in stereo using two or three microphones. Typical placement for pop/rock recording is shown in Figure 14–9. In a jazz context the microphones would normally be placed slightly outside the rim of the instrument, giving just a bit more distance to the instrument.

When picked up at such short distances the piano may sound quite bright, and the hammers may have to be softened by a piano technician. Any slight mechanical problem may become very apparent with close microphone placement, and the technician should be prepared to fix such problems. Keep in mind that any changes made to the hammers (hardening or softening) cannot always be undone easily or quickly. The technician and player should be in agreement on any changes made to the instrument.

Most engineers favor cardioid microphones for close placement, since the proximity LF rise will usually enhance the sound of instrument. Microphones placed outside the rim may be omnidirectional or cardioid, at the choice of the engineer or producer. Some engineers prefer to use boundary layer microphones, fastening them to the underside of the piano cover using double-sided tape. This minimizes the effect of reflections inside the instrument.

As an aid to minimizing studio leakage into the piano microphones, many engineers position the piano so that its cover, when open, faces away from the other players, thus shielding the instrument from direct sounds in the studio. If more isolation is required, the piano cover may be positioned on "half-stick" and a blanket placed over the opening, as shown in Figure 14–10.

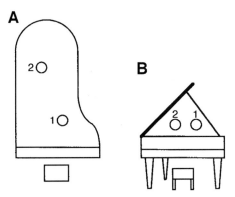

Pan 1 left of center and 2 right of center for normal stereo perspective. Both microphones about 0.3 meter (1 ft) above strings.

FIGURE 14-9 ———

Recording the piano: top view (A); front view (B).

While the basic piano tracks should be recorded dry, reverberation will always be added in postproduction. The presentation on the stereo soundstage is normally in stereo; however, if the mix is a hard-driving, complex one, the piano may sound clearer if mixed to mono and positioned where desired.

Blanket covering opening; piano cover on half-stick

FIGURE 14-10

Recording the piano with added isolation.

RECORDING THE ACOUSTIC BASS

There are several methods for recording the bass:

1. Microphone on a floor stand
2. Nesting a microphone behind the tailpiece of the instrument
3. Microphone at bass amplifier/loudspeaker unit
4. Direct-output from instrument pickup into the console

Using a floor stand, as shown in Figure 14–11, a cardioid microphone can be precisely placed to pick up the desired combination of fundamental sound along with the sounds of the player's fingers on the strings. For acoustical jazz it is important to achieve this, and we recommend this approach as the best overall choice. It may be helpful to place a set of 1 m (40 in) high baffles behind the microphone to reduce leakage from other instruments.

Some engineers prefer to place a small format omnidirectional microphone in the space between the tailpiece and the body of the instrument. It is usually wedged in place using a piece of foam rubber, as shown in Figure 14–12. The advantage as here is that any movements of the instrument will not alter the microphone-to-instrument distance; a disadvantage is that the pickup of finger articulation will be minimized.

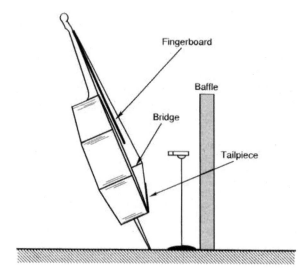

Fingerboard

Baffle

Bridge

Tailpiece

FIGURE 14-11

Recording the bass with a stand-mounted microphone.

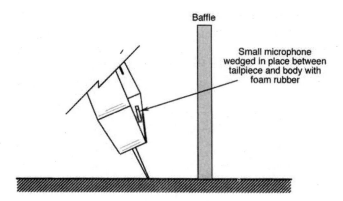

FIGURE 14-12 ————

Recording the bass with a microphone mounted on the instrument.

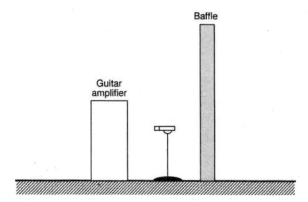

FIGURE 14-13 ————

Picking up the bass acoustically from the bass amplifier.

Under some adverse conditions the engineer will have no other recourse than to place a microphone directly in front of the loudspeaker in the bass amplifier/loudspeaker unit, as shown in Figure 14-13. Many times, a quick stage setup or change-over will not provide the time necessary to work out the details of a better pickup method. Many bass amplifiers are noisy and may be distorted at high output levels, both of which can cause problems in postproduction. However, if the amp is clean and noise-free, you can get a good recording this way.

A direct console feed from the instrument's pickup is an excellent way to achieve a good bass sound with no leakage whatever. Be sure that you have a good active splitter box that will allow you to sample the signal directly out of the bass, while sending an uninterrupted signal on to the bass amplifier. Details are shown in Figure 14-14A and B. Most direct output feeds are unbalanced (see Chapter 8 under Microphone Splitters) and should be fed into a console input set for line input operation. Phantom powering should be turned off.

Given the choice, most engineers will want to lay down at least two bass tracks, one direct and the other via a microphone pickup. The two will often be combined in postproduction. (Be on the lookout for anti-phase cancellation when the signals are combined later on.)

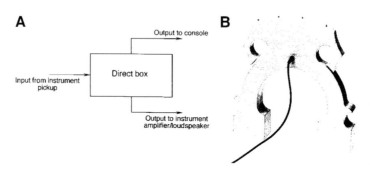

FIGURE 14-14 ————

Recording the bass: using
the direct output from the
bass (A); a pickup mounted
on the instrument's bridge
(B). (Photo courtesy of
AKG Acoustics.)

CHOICE OF MICROPHONES FOR BASS PICKUP

Bass tracks are invariably compressed and equalized during postproduction. While much in the way of equalization can be done at that time,
it is always best to get the ideal sound on the basic tracks in the first
place. There are a number of excellent dynamic microphones for bass
pickup; these are often models that have been designed with a slight bass
rise and with adequate headroom at low frequencies. Many engineers
prefer to use large diameter Braunmühl-Weber capacitor types, since
these models often have distinctive high frequency signatures which both
engineer and performer may deem desirable.

VOCALS AND VOCAL GROUPS

In performance, most vocalists make use of handheld "vocal microphones"
whose response characteristics are internally equalized to produce a fairly
smooth response at close operating distances. In the studio, the vocalist is
normally recorded via a Braunmühl-Weber dual diaphragm large format
capacitor model. Years of acclimation have convinced most vocalists as
well as engineers that this is the only way to go.

There is much to be said in favor of this microphone type: the models are all slightly different in response, and the engineer can usually pick
one that truly enhances the performer's voice at high frequencies. As a
group they are robust and can take high vocal levels in stride. Many
older tube models dating from the early 1960s are highly regarded and
are routinely used for these purposes. (See Chapter 21 for a discussion of
older classic microphone models.)

The normal studio setup is as shown in Figure 14–15. The microphone, set in cardioid position, is placed in front and above the vocalist,
with room for a music stand. The operating distance is approximately
0.6 m (24 in). A sheer pop screen is normally positioned in front of the
microphone as shown in order to control inadvertent pops of wind from
the vocalist. Be sure to angle the music stand so that there are no direct
reflections from the vocalist to the microphone. It is customary to provide
a stool for the vocalist, whether or not it is actually used. In a tracking

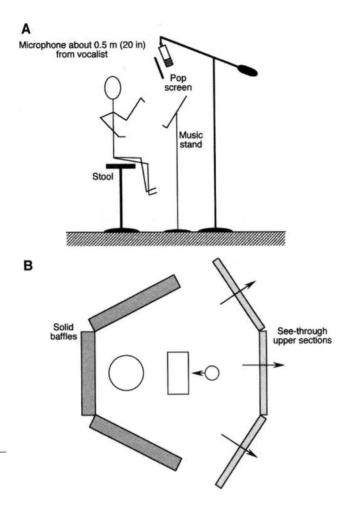

A

Microphone about 0.5 m (20 in)
from vocalist

Pop
screen

Music
stand

Stool

B

Solid
baffles

See-through
upper sections

FIGURE 14–15 ————

Recording a vocalist: side
view (A); top view (B).

session, the vocalist will normally be surrounded by baffles on three sides, and in extreme cases of studio leakage it may be necessary for the vocalist to perform in a vocal booth. This is a rather confining environment and should be chosen as last resort.

Wherever the vocalist is located, it is important that headphone monitoring be carefully tailored to the vocalist's tastes. Typically, the vocalist will want to monitor a stereo mix with a generous amount of the vocal track itself, complete with stereo reverberation. In the unlikely event that you are using a compressor on the vocal track going to tape, *do not* put that signal in the vocalist's monitor mix.

There are so many important details here that it is strongly recommended that you cover as many points in setup as you can before the vocalist arrives for the session. You can always find someone on the studio staff who will be glad to help you set up. The psychology of the moment is critical; nothing is more comforting to a vocalist than hearing a truly clean and clear sound the first moment the headphones are put on.

AFTER THE INITIAL TRACKING SESSIONS

On many occasions the vocalist will want to record basic vocal tracks after the initial instrumental tracking sessions have been finished. In this case there will be few, if any, performers in the studio, and the isolation measures we have discussed will probably not be necessary.

Be prepared to do vocal insert recording efficiently and smoothly, if it is required. Make sure you have an experienced tape operator who can keep accurate notes on the take sheets.

As vocal tracking gets underway the singer will want to come into the control room to listen to the tracks that have just been recorded. Make sure that you have a reasonable monitor mix up and running for the singer and other artists as soon as they get to the control room.

While many of the points we are making here have nothing to do with microphone technique as such, they are very important to the smooth running of any pop recording session – and directly reflect on your general competence and microphone choices.

RECORDING BACKUP VOCALISTS

You can use as many recording channels as you have to spare, but a minimum of two may suffice, as shown in Figure 14–16. Since more than one singer will be on each microphone, do not hesitate to reposition the singers slightly in order to correct balances. Experienced backup singers are always aware of this and will gladly follow your requests. If you find it necessary to compress the stereo signals, make sure that you use two compressors that are stereo-coupled in order to maintain overall balances. A backup chorus, if needed, can best be recorded using the classical techniques discussed in Chapter 13.

RECORDING THE GUITAR

The acoustic guitar is equally at home in classical as well as popular music of all kinds. While intimately associated with Spain, it has become

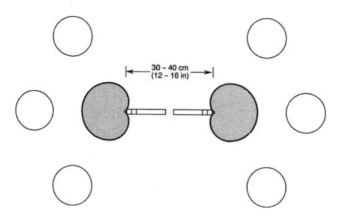

FIGURE 14–16 ————

Recording a backup vocal group in stereo, top view; circles show approximate positions of singers.

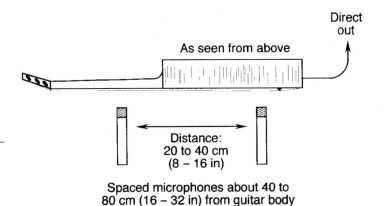

FIGURE 14-17

Recording the acoustical guitar with microphones and direct input to the console.

a truly universal instrument. For most of its existence it survived only as an acoustic instrument; however, during the last five decades it has emerged in a dual role in both acoustical and amplified form. For rock applications, the solid body guitar has become the chief instrumental musical element. With its solid body structure, the only resonance present in the instrument lies in the strings, and the sound we hear is that of the strings alone suitably amplified and processed.

In the studio, the acoustical guitar is normally recorded both with microphones and via direct electrical output; in many cases a final mix will make use of both inputs. In this section we will consider both options.

Figure 14–17 shows a conventional stereo method for recording the guitar acoustically. Most engineers prefer spaced microphones as shown here, but a coincident pair of cardioids is also an option.

When using spaced microphones, if the guitar is to appear in mono in subsequent postproduction, it is best if only one of the recorded channels is used, suitably equalizing it for the desired spectral balance. Alternatively, a coincident pair can be mixed directly to mono with no phase cancellation problems.

There is much added flexibility in postproduction if a third track is used to record the direct output from the guitar. The timbral difference between the direct and microphone signals will be considerable, and this can be used to create a fairly widely spaced stereo presentation.

Modern guitar amplifiers have stereo outputs, and when recording the solid body instrument it would be a mistake not to record both outputs for the maximum postproduction flexibility.

SYNTHESIZERS

Like the solid body guitar, the synthesizer ("synth", as it is called) can only be recorded directly from a stereo pair of outputs. In many cases synth players will want to lay down additional tracks by overdubbing, so manage your track capability accordingly.

WOODWIND AND BRASS INSTRUMENTS

As noted in Chapter 13, woodwind instruments have fairly complicated radiation patterns, and this dictates that microphones should not be placed too close to them if a balanced sound is desired. Often on television performances we often see small microphones clipped to the bells of clarinets, primarily as a matter of expedience. If Figure 13-2 is any indicator, the resulting sound is likely to be very bass heavy, and considerable equalization will be needed to create a good timbre. Brass instruments will fare somewhat better with such close pickup, since all radiation is by way of the bell. Figures 14–18 and 14–19, respectively, show how individual woodwind and brass instruments may be recorded close-in for best balance. The dimensions shown in Figures 14–18 and 14–19 represent minimum values; you can always place the microphone farther away, consistent with leakage problems in the studio. See Meyer (1978), Dickreiter (1989) and Eargle (1995) for additional information on the directional characteristics of musical instruments.

The French horn is a special case. It is normally heard via room reflections, since the bell always fires to the rear of the player. If more control is needed it is best to place a microphone oriented overhead at 90° to the bell axis. In that position the microphone will pick up some of the "buzz" that characterizes the raw sound of the instrument, giving the recorded sound additional presence. Whenever possible, a set of reflective baffles should be placed behind the instruments.

STRING INSTRUMENTS

Data given in Figure 13–3 show the complex nature of radiation from bowed string instruments, indicating how difficult it is to obtain a natural

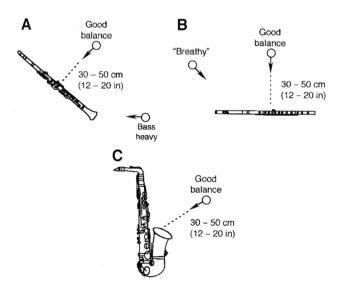

FIGURE 14-18

Recording woodwind instruments: flute (A); oboe/clarinet (B); saxophone (C).

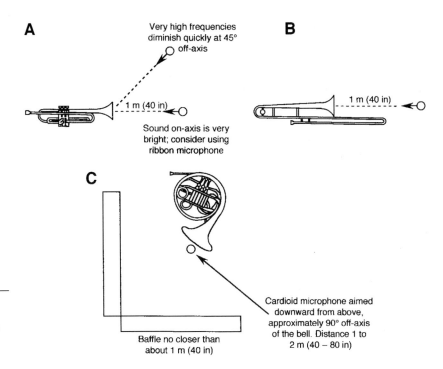

FIGURE 14-19

Recording brass instruments: trumpet (A); trombone (B); French horn (C).

balance at close operating distances. Nevertheless, clip-on microphones are often used on-stage when string instruments are used in combination with much louder instruments. Since the bowed instruments are normally used in multiples, we will discuss their pickup below under The Large Studio Orchestra.

The harp can be recorded as shown in Figure 14–20. A spaced stereo pair of omnidirectional microphones will produce the best timbre, but in a crowded studio it may be mandatory to use cardioids for added isolation.

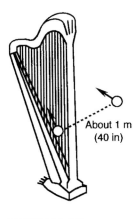

FIGURE 14-20

Recording the harp in stereo.

ENSEMBLES IN THE STUDIO

Our discussion here will be limited to three jazz ensembles ranging from small to large and a single large studio orchestra, such as might be used for film scoring.

A JAZZ TRIO

Basic layout and panning

Figure 14–21A shows the studio layout for a basic jazz instrumental trio consisting of piano, drums and bass. The simplicity of this ensemble allows us to explore the basic advantages of stereo recording of both the piano and drum set.

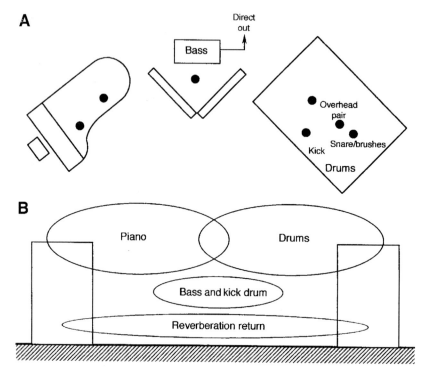

FIGURE 14-21

Recording a jazz trio: studio layout (A); target stereo soundstage (B).

Whenever possible, the engineer and producer should place the musicians in the studio with the same left-to-right positioning as intended in the final postproduction mix. The reason here is that any studio leakage will be between adjacent instruments as they appear on the stereo soundstage and thus not create any conflicting spatial cues. Studio leakage is not necessarily a bad thing, and it can vitally enhance a performance rich in acoustical cues when the instruments are so tightly clustered together.

The piano has been placed at the left, since the open cover of the instrument will reflect sound to both the bassist and drummer. The bassist is placed in the center with both microphone and direct pickup. The drum set is placed at the right and uses an overhead stereo pair with additional microphones on both the snare and kick drums. The primary purpose of the snare microphone is to pick up the delicate playing of brushes on the snare drum.

Establishing the monitor mix

Both producer and engineer must be in mutual agreement on the stereo layout of the recording. Here, the aim is to have the piano be reproduced from the left half of the stereo stage, and this calls for panning the treble piano microphone at the left and the bass piano microphone at the center. Because of the considerable leakage between the signals at both microphones the sound will appear from the left half of the stereo stage, with high frequencies tending toward the left and middle and low frequencies tending toward the center.

The bass microphone and direct pickup will both be panned to the center in whatever proportion best fits the music. The balance between the two can in fact vary throughout the recording from tune to tune.

The kick drum microphone will be panned to the center, and the drum overheads will be panned center and right. The microphone on the snare will be subtly mixed in and panned between center and right.

In a typical studio environment, the piano and bass inputs will be fed to the stereo inputs of an external reverberation generator whose stereo returns will be panned to the left and right signal buses. The choice of whether to use any artificial reverberation on the drum microphones is open for consideration. Normally none is used.

The reverberation parameters for this recording would normally be set for a LF (below 500 Hz) reverberation time of about 1 second, with reverberation time above that frequency slightly longer.

The target stereo soundstage

The representation at Figure 14–21B gives a picture of the resulting soundstage as intended by the producer and engineer. Note that the stereo presentation is wide, yet preserves good center-stage imaging of the bass as well as portions of the piano and drum set. The reverberation return signals should be panned full left and right in order to simulate the widest spatial separation. For a group as small as this, and with instruments that are equally balanced acoustically, there may be virtually no need for headphone monitoring.

JAZZ VOCAL WITH SMALL INSTRUMENTAL GROUP

Basic layout and panning

Figure 14–22A shows the basic studio setup and panning assignments for a jazz group consisting of vocalist, two solo saxophones and a full rhythm section. As before, the studio left-to-right setup is in keeping with the anticipated sound stage.

The piano has been replaced by a Hammond electronic organ, a mainstay in blues-type jazz vocals. The organ is recorded in stereo with microphones next to its HF and LF loudspeakers. The Hammond B-3 model is never recorded direct-in because of the extensive pre-processing of the signal and the fact that it normally uses the Leslie loudspeaker system, with its characteristic HF rotating loudspeaker array. The acoustical guitar is picked up with both microphone and direct output. Both organ and guitar are arrayed in stereo, respectively on the left and right, enhancing the musical dialog that often takes place between them.

The two saxophones are placed left and right. When they are playing in counterpoint or dialog, their left and right panned positions are

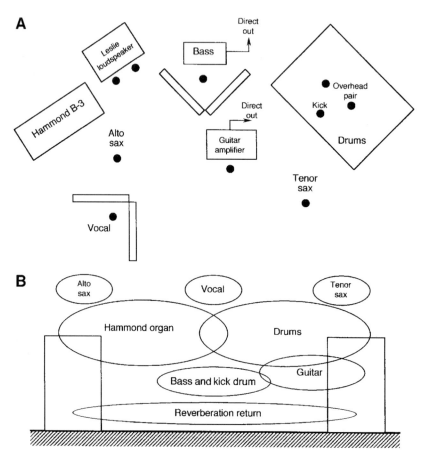

A

FIGURE 14–22

Recording jazz vocal with small instrumental group: studio setup (A); target stereo soundstage (B).

maintained; when either instrument is playing a solo chorus it may be panned to the center of the stereo stage. The reason here is purely tradition and expectation. Bass and drums are positioned as in the previous example.

Establishing the monitor mix

The first step is to establish a basic balance with bass, drums and organ; listen for a smooth but detailed soundstage from left to right. Then the guitar can be introduced into the mix. As a final step, balances among the vocalist and saxophone soloists are made. Reverberation sends should be taken from all instruments, except possibly the drums, and returned to the monitor mix with full left and right panning.

The target stereo soundstage

The representation in Figure 14–22B shows the stereo soundstage as intended by the producer and engineer. There are three basic spatial layers in the recording: front-most are the three essential soloists. The secondary layer includes the rhythm elements, and the reverberation return constitutes the back layer.

At times, the rhythm elements will take on primary importance, and when the do they should be boosted in level according to their importance. A basic skill of any recording engineer is knowing when – and how much – to alter the level of an instrument that moves from one layer to another. An experienced producer is of great help here.

Always bear in mind that not all elements in the mix can have equal importance at the same time. Both spatial and frequency dimensions must be considered; whenever possible, important elements in the mix should be perceived as coming from separate directions. Likewise, the mix, as perceived overall, should exhibit a uniform spectrum from top to bottom.

JAZZ BIG BAND

Basic layout and panning

The standard large jazz band normally consists of four trombones, four trumpets, and five saxophones in the brass and wind departments. Rhythm elements consist of drums, bass, piano and guitar. Brass and wind players may also double, respectively, on French horn, clarinet or flute. There may be an additional percussion player as well. Figure 14–23A shows a typical studio layout. The trumpets are usually placed on a riser

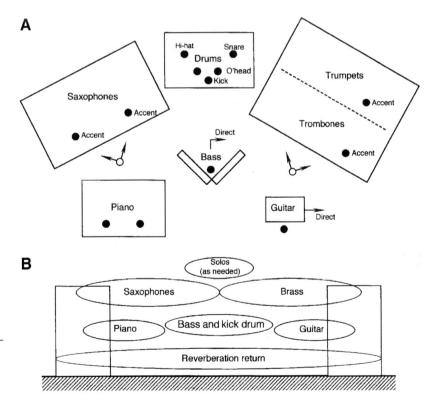

FIGURE 14-23

Recording a jazz big band: studio layout (A); target stereo soundstage (B).

behind the trombones, and it is not unusual for the various wind sections of the band to stand when playing as an ensemble. Stereo pickup of both saxophones and brass is typical, with accent microphones in place for solos. The engineer needs accurate cueing from the producer for spotting the solos. Usually, a single stereo reverberation system will adequately take care of all ensemble demands for ambience. However, a vocal soloist would certainly call for an additional reverberation system, most likely with shorter reverberation time settings.

Target stereo soundstage

Both saxophones and brass are panned across the stereo soundstage to form a continuous presentation. The spotting of solos, depending on how extended they may be, are panned into the center. The rhythm elements form another layer, arrayed in stereo with piano at the left, drums in the center, and guitar at the right. The overall soundstage is as shown in Figure 14–23B.

In order to avoid a cluttered setup, most engineers would opt for stereo microphones for the basic stereo pickup of the saxophones and the brass. Pickup of the rhythm elements would be in accordance with the suggestions made earlier in this chapter. The safest approach to recording a group this complex would be to assign each microphone to its own recording track, while at the same time making a two-track stereo monitor mix. There are a number of engineers who are truly adept at making "live to two-track" mixes of big bands. An automated console, with adequate subgrouping of inputs, makes the job a bit easier.

THE LARGE STUDIO ORCHESTRA

As discussed here, the large studio orchestra is typical of the ensemble that would be used in motion picture sound track scoring. Superficially, the group may resemble a small symphony orchestra. The main pickup will in fact look a great deal like those discussed in Chapter 13.

The engineer must determine, in consultation with the composer/ arranger and the recording producer, the requirements of the score. Any instrument that will be featured in a solo role, however momentary, will probably need an accent microphone. The reason for this is that post-production mixing of music tracks behind dialog and/or effects will ordinarily require that individual solo instruments be highlighted in order to be heard effectively in the final film mix. This is a judgment that cannot readily be made during scoring sessions.

However, the tracks laid down for opening and closing credits may well be used exactly as balanced by the mixing engineer during the scoring sessions. The only new requirement introduced here is that the mix will, in today's creative environment, be done in five-channel format, three in front plus two surround channels, for motion picture purposes.

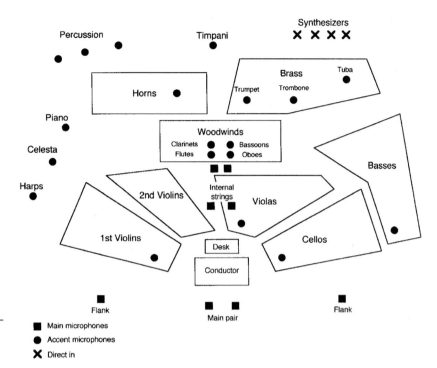

FIGURE 14-24

Studio layout for a large studio orchestra.

The string ensemble, as in classical recording, always sounds best when the microphones are no closer than absolutely necessary to delineate them with respect to the winds and brasses. Many a great score has been ruined by too-close string microphone placement, with its characteristic "screech." Figure 14–24 shows a typical studio layout for a large scoring session. Note that the strings are picked up primarily via the main four microphones across the front and the internal overhead pair on the inner strings. The accent microphones are used only for highlighting solos, and they are often implemented only during postproduction sessions.

MODERN STUDIO ACOUSTICS AND CONTROL OF LEAKAGE

Half a century ago, recording studios tended to be fairly dry acoustical environments in order to meet needs growing out of broadcasting and film sound recording, with their limited dynamic ranges. Over the decades we have seen more varied spaces as the recording arts have evolved. Today's recording studio is apt to have a number of environments within a larger space. For example, one end of the studio may be more acoustically live than the other; or perhaps the entire studio will be covered with movable wall sections that can be turned inward or out to create local live or damped playing areas, as shown in Figure 14–25.

A

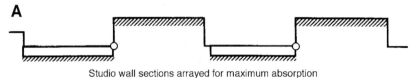

Studio wall sections arrayed for maximum absorption

B

Studio wall sections arrayed for minimum absorption

FIGURE 14–25

Live and absorptive areas in a modern studio.

A

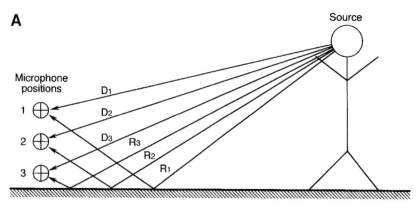

B

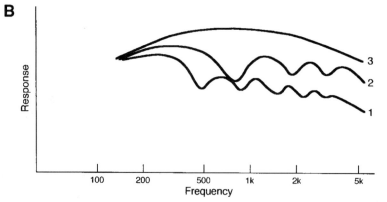

FIGURE 14–26

Reflections in the studio: microphone positions (A); response (B).

Modern studios will have several isolation areas. These are used primarily for vocals and any soft instrumental elements in the mix. Some studios have a large space, virtually a second studio, adjacent to the main area and large enough to contain a sizeable string group. Large sliding glass doors provide the necessary sight lines into the main studio.

Discrete reflections in the studio can be problematic, as shown in Figure 14–26A. For example, in recording cellos and basses, the microphone may be fairly close to the floor. If the microphone is about

FIGURE 14-27 ————

A boundary layer microphone. (Photo courtesy of Crown International.)

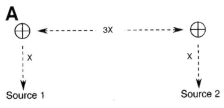

A

Source 1 ←------ 3X ------→ Source 2

X   X

Both sources are equal

FIGURE 14-28 ————

The "three-to-one" rule in omnidirectional microphone placement: both sources about equal in level (A); source 1 louder than source 2 (B).

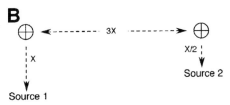

B

←-------- 3X --------→

X   X/2

Source 2

Source 1

Source 1 is louder than Source 2

equidistant from the sound source and the floor, as shown at position 1, the reflected path can create an irregularity in response as shown at B. Progressively moving the microphone toward the floor will minimize the response irregularities.

In most cases it is best to place the microphone directly on the floor and avoid interferences completely. *Boundary layer* microphones are specifically designed for this application. A typical model is shown in Figure 14–27. Normally, an omnidirectional pattern is chosen, but directional boundary layer microphones can be used if more separation is required.

THE THREE-TO-ONE RULE

Another problem arises in smaller studios where players are often seated closer together than is desirable. Figure 14–28A shows what is called the "three-to-one" rule. That rule states that, when using omni microphones, the distance from a microphone to its target sound source should be no greater than one-third the distance from that microphone to the nearest interfering source. The three-to-one ratio will reduce the level from the interfering source by an average of 10 dB, which is quite acceptable. When one instrument is significantly louder than the other, then the adjustment shown at B must be made. Obviously, using cardioid microphones will diminish interference problems all the more.

CHAPTER 15

SURROUND SOUND MICROPHONE TECHNOLOGY

INTRODUCTION

While two-channel stereo has been the mainstay of consumer sound reproduction for nearly a half-century, surround sound as originally developed for motion pictures has technical roots that go back even earlier. For the consumer, surround sound as a central requirement in music-only presentation in the home has had a spotty history. Quadraphonic (four-channel) sound was introduced in the mid-1970s and failed, chiefly because the technology proposed for it was not sufficiently developed. During the mid-1990s, surround sound was reintroduced to the consumer as an integral part of the home theater revolution, with its five-channel loudspeaker array consisting of 3 loudspeakers in front and 2 at the sides slightly to the rear of the listener. The basic plan was patterned after the normal loudspeaker setup in motion picture theaters of the day. A primary performance benefit of the "new" video-based surround sound was the use of a front center channel, which anchored center-stage events accurately in that position – regardless of where the listener was located. The added benefit of global ambience as fleshed out by the back channels was equally beneficial.

The earliest five-channel carriers for music-only surround programming include the DVD Video format (with Dolby AC-3 encoding) and the DTS (Digital Theater Sound) CD format, both presenting five full-range channels and a single subwoofer effects channel operating below 100 Hz. (For obvious reasons, the format is known as 5.1 surround sound.) As the new millennium got under way, two formats, DVD Audio and Sony/Philips SACD (Super Audio Compact Disc), were both proposed for audio-only applications for surround sound. At the present time, both these mediums have made a respectable bid for marketplace acceptance, but neither has been the runaway success that many had expected.

The first five years of the new century have seen so much development in surround technology, including microphone design, microphone pickup techniques, and playback options, that it will be advantageous to discuss the technology first in a chapter devoted solely to those topics. The next chapter will then cover case studies of the various techniques.

For our purposes we will consider four categories of surround sound recording and playback:

1. *Stereo-derived:* There are many formats here, and they are basically derivations of current stereo techniques in that they make use of both real and phantom images as well as ambient effects that arise from decorrelated multichannel sources. Quadraphonic and current motion picture techniques fall under this category, as do special formats such as the TMH Corporation's 10.2 configuration. Virtually all of the surround remixes of both legacy and modern multitrack source tapes fall in this category.

2. *Single-point pickup:* These techniques are used to sample and play back a global three-dimensional sound field. The British Soundfield microphone is an early example here using first-order directional patterns. The more recent Eigenmike makes use of higher order patterns. In either cases there is a direct correspondence between the pickup and playback directions, and the general aim of these systems is to duplicate natural spatial cues in the playback environment. The number of pickup elements is normally limited by the order of the microphone patterns, but the playback setup can usually handle more loudspeaker sources than microphones if their drive signals are properly derived. In all of these systems it is essential that the listener be positioned at the center, or "sweet spot", of the loudspeaker array.

3. *Transaural, or head-related, pickup:* This spatial sound transmission technique uses a fairly small number of loudspeakers to duplicate at the listeners' ears the exact amplitudes and time relationships that were physically present in the pickup environment. It is very critically dependent on loudspeaker–listener positioning, and its major use is in controlled listening environments (such as those afforded by computer workstations) where the listener's head is confined to a single position.

4. *Systems with parallax:* Like an acoustical hologram, these systems allow the listener to move about in the listening space while virtual source locations remain stationary. Generally, a relatively small number of recorded tracks may be involved, and the real complexity comes in playback signal processing, where positional information is superimposed on the various sources via impulse measurements and signal convolution. The technique is experimental at present, but it holds great promise for use in special venues.

For each category summarized above we will describe the general technical requirements for both recording and optimum playback.

STEREO-DERIVED SYSTEMS

WHAT IS AMBIENCE?

There are three acoustical elements necessary for convincing surround sound reproduction:

1. Accurate pickup of direct sound originating on-stage; soundstage imaging should be natural and unambiguous.

2. Pickup of sufficient early reflections from the stage and front portion of the recording space to convey a sense of room size and dimension. These signals normally fall in the range from 25 to 60 ms after the receipt of direct sound and are generally conveyed by all of the loudspeakers in the surround array.

3. Pickup of uncorrelated reverberation and its subsequent presentation over the entire loudspeaker array. The onset of the reverberant field normally occurs at about 80 ms after the receipt of direct sound.

Conventional two-channel stereo can deliver the frontal sound-stage quite accurately for a listener seated directly on-axis, but cues pertaining to early reflections and reverberation are limited by the relatively narrow presentation angle in the playback environment. We need to determine how many channels are actually necessary to do justice to the demands of surround sound. Tohyama et al. (1995) developed data showing that a minimum of four channels is necessary to create an accurate enveloping sound field in the consumer listening environment. The basis for their study is shown in Figure 15–1. In this figure, the inset shows an artificial head located in the completely

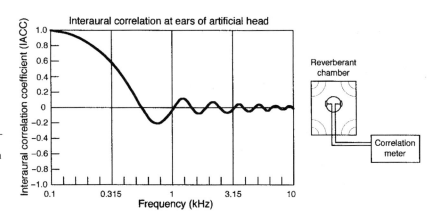

FIGURE 15–1

Interaural cross-correlation (IACC) between the ears of an artificial head placed in a reverberant field.

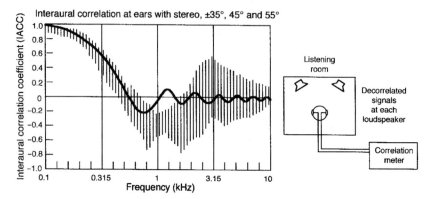

FIGURE 15–2

IACC for stereo playback of uncorrelated reverberation in a typical listening space.

diffuse sound field of a reverberation chamber, which is analogous to the acoustical field component in a concert hall which conveys the normal sense of envelopment. The test signal consists of a slowly swept band of noise ranging from 100 Hz to 10 kHz. The signals reaching the microphones at the ear positions in the artificial head are compared, and the mathematical cross-correlation between the ear positions is measured and shown in the graph. At very low frequencies the correlation is unity, inasmuch as the microphone spacing is small relative to the received wavelength.

As the signal increases in frequency the cross-correlation converges to an average of zero, indicating that the received signals at the artificial head are essentially uncorrelated, a basic condition for conveying a sense of spatial envelopment at the ears of the listener.

Moving on to Figure 15–2, we now take an uncorrelated two-channel stereo recording made in the diffuse field of Figure 15–1 and play it back in a typical living room environment. The artificial head is again used to pick up the swept signal, and the interaural crosscorrelation is again measured and plotted, averaging three sets of stereo loudspeaker playback angles. The graphed data show that the measured signal correlation at the ears is not uniform, and in fact produces a significant compromised sense of envelopment, especially at the critical midrange frequencies between 800 Hz and 3150 Hz. As good as it sounds, stereo is not capable of generating a convincing sense of spatiality around the listener.

Figure 15–3 shows similar measurements, this time using four uncorrelated playback channels, averaging them at six different sets of bearing angles around the measurement system. It is clear that the four-channel array very nearly produces the same spatial cues as the reference condition shown in Figure 15–1, indicating that four (or more) channels of ambience information can successfully produce an accurate enveloping sound field in the home listening environment. A number of multichannel reverberation systems currently produce such decorrelated sound fields.

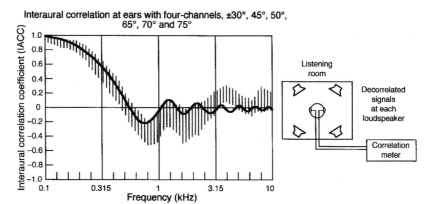

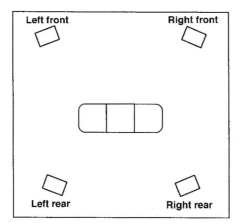

FIGURE 15-3

IACC for quadraphonic playback of uncorrelated reverberation in a typical listening space.

FIGURE 15-4

A typical playback setup for quadraphonic sound reproduction in the home environment.

DETAILS OF THE PLAYBACK SETUP

Figure 15–4 shows for reference purposes a typical quadraphonic playback arrangement. While the system could produce a good impression of recorded ambience, the very wide spacing of the front pair of loudspeakers (90°) was disturbing to many listeners, primarily because it did not reproduce front-center phantom images convincingly. Because of this, many consumers routinely reduced the frontal listening angle to about 60°, while leaving the back angle in the range of 90°.

Surround sound for motion pictures has evolved over the years to the arrangements shown in Figure 15–5. There are either two surround channels or three, depending on the vintage of the installation and its implementation of the Dolby EX-Plus center-rear surround channel. In the motion picture theater, multiple surround loudspeakers are always used in order to produce highly uncorrelated signals, the specific source of which patrons cannot readily identify. This is a very desirable condition and normally fits the use of the surround channels as carriers of special environmental, often large-scale, effects in the typical motion picture.

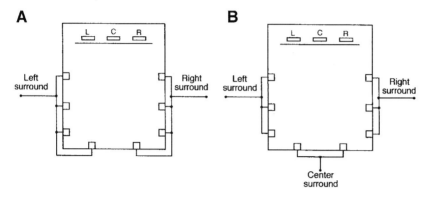

FIGURE 15-5

Loudspeakers in the motion picture theater: using two surround channels (A); using three surround channels (B).

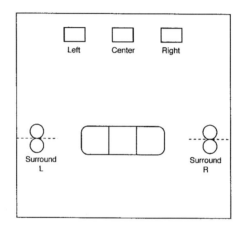

FIGURE 15-6

Typical home theater surround sound loudspeaker locations.

It is difficult to transport the complexity of the motion picture approach into the home environment, and more often than not, only a pair of loudspeakers will be used to convey a sense of surround envelopment for on-screen effects. A typical home theater loudspeaker array is shown in Figure 15–6. Here, *dipole* loudspeakers are used for the surround channels. The dipole loudspeaker exhibits deep nulls in its response at angles of 90° relative to forward and rear axes, and the side null axes are aimed at the primary listening area. The result is that the signal from the surround channels reaches the listeners primarily by way of room reflections and is thus less likely to call attention to itself as coming from distinct positions left and right of the listeners.

Dipole loudspeakers work well for surround music channels that are primarily reverberant in their signal content. For more generalized music presentation, the playback format shown in Figure 15–7 is recommended. This figure illustrates the standard loudspeaker configuration recommended by the ITU (International Telecommunications Union, 1994) for the setup of listening environments for the mixing of surround sound programs for professional activities. While these recommendations

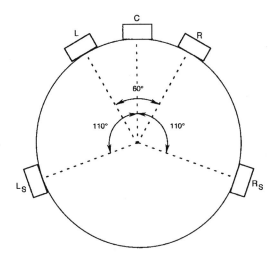

FIGURE 15-7

ITU recommended reference loudspeaker locations for surround sound monitoring; the nominal ±110° positions may vary between 100° and 120°.

are appropriate for surround sound presentation, it is felt by many workers in the field that the left and right front loudspeakers, with their target spacing of 60°, may be too widely spaced for normal stereo listening in the home.

In laying out a surround sound mix, the three frontal channels are important in defining and positioning of primary musical elements. Our hearing capability can lock on elements in the frontal soundstage and identify their lateral position within a few degrees. This acuity is far more stable with three frontal channels than it is using phantom images in two-channel stereo, and it substantially enlarges the effective seating area for all listeners.

While we are very aware of sounds originating from the sides and rear, we have considerable difficulty in determining their actual positions. Sounds arriving from the sides are generally highly uncorrelated at the ears and as such contribute to a sense of ambience in the reproduced recording. Because of the ear's difficulty in assigning specific directionality at the sides, a single pair of channels can handle the needs for conveying ambience very well, especially if the loudspeakers are arrayed in large quantities, as in the motion picture theater, or as diffuse-radiating dipole loudspeakers. Thus, the obvious use of the surround channels in classical music is in conveying early reflections and reverberant sound in the recording space. Note that the rear loudspeakers in the ITU configuration are positioned more to the sides than the back. In this position they can convincingly deliver both early reflections and reverberant sound.

Pop mixes are routinely made with secondary musical elements panned into the surround channels; here we include such elements as background vocals and rhythm fills. Live concerts usually fare very well in surround sound through the combination of audience reactions and secondary musical elements in the rear channels.

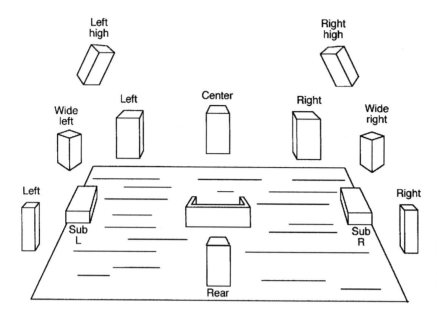

FIGURE 15–8

Layout of the TMH
Corporation proposed
10.2 array.

Extending the 5.1 concept to a greater degree, Holman (2000) has suggested the 10.2 channel approach shown in Figure 15–8. The wide-front channels are used primarily to enhance important early side reflections, which help to define the acoustical nature of a good recording space; the center back channel ensures spatial continuity and permits unambiguous rear imaging. The front overhead channels provide height information which, like the front-wide channels, enhances spatiality and adds realism to the presentation. The use of two low frequency channels restores some of the spatial subtlety in the lower frequency range which results from side-to-side pressure gradients at long wavelengths. Such effects may be lacking when a single channel is used below 100 Hz. Holman further points out that the exact positions of the added loud-speakers are subject to a moderate amount of leeway in actual positioning in the playback space. In practical application of the 10.2 format, the notion of "virtual" microphones (see Chapter 19) is useful in the generation of secondary reflected and ambient signals.

MICROPHONES AND ARRAYS FOR STEREO-DERIVED SURROUND SOUND

While most surround sound recording is carried out using conventional microphones and techniques, with assignment of the various signals into the five cardinal points of a surround sound array, a number of microphone designs and microphone arrays are used for recordings intended primarily for surround presentation. Some of them are discussed below.

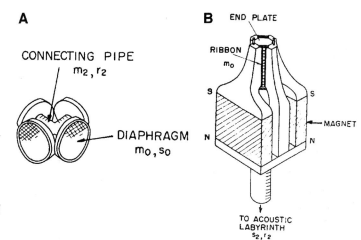

A

CONNECTING PIPE
m_2, r_2

DIAPHRAGM
m_0, s_0

B END PLATE

RIBBON
m_0

MAGNET

TO ACOUSTIC
LABYRINTH
s_2, r_2

FIGURE 15-9

Single-point quadraphonic microphones: using capacitor elements (A); using ribbon elements (B). (Figure courtesy of *Journal of the Audio Engineering Society*.)

COINCIDENT ARRAYS

Going back to the quadraphonic era, a number of microphones were developed that provided for one-point pickup of sound from four directions, a concept that is still useful in certain aspects of surround sound. Typical here is the single-point four-cardioid array developed by Yamamoto (1975), shown in Figure 15–9, either as a group of cardioid capacitor elements (A) or ribbon elements (B) picking up sound in the azimuthal (horizontal) plane. In either case, the mechanical and acoustical elements in the designs have been adjusted so that the target cardioid directional characteristics of each element are maintained around the circle with smooth transitions between adjacent pairs.

NEAR-COINCIDENT ARRAYS

The Schoeps KFM 360 spherical array

Developed by Bruck (1997), the KFM 360 produces a dual set of MS stereo patterns, one set facing the front and the other facing the rear. Both MS sets are attached to a sphere 18 cm in diameter, as shown in Figure 15–10A. The pattern choices are shown at B. The user can choose the forward patterns independently of the rear patterns, thereby altering front-back perspectives. The microphone array is fed to the DSP-4 control unit shown in Figure 15–11A, and a signal flow diagram for the unit is shown at B.

The gradient microphones enable this fairly small array to produce excellent fore-aft spatial delineation, and the effect can be further heightened through use of the rear channel delay function. Note that the control unit produces a front-center output via the Gerzon 2-to-3 matrix circuit (see below under Systems with Parallax).

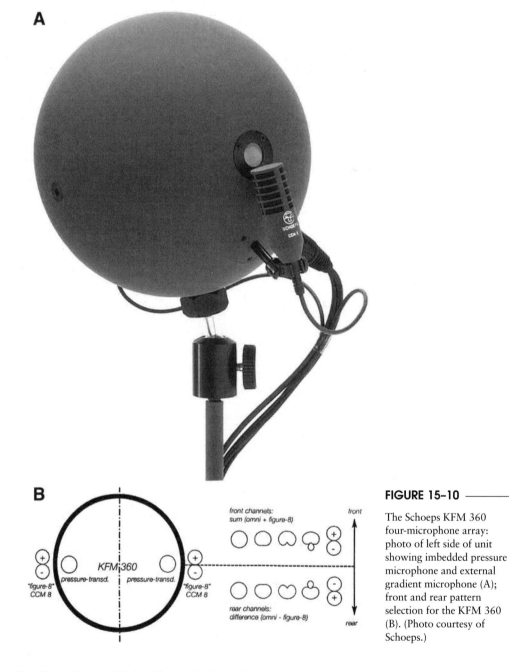

A

B

KFM 360

pressure-transd.   pressure-transd.

"figure-8"   "figure-8"
CCM 8   CCM 8

front channels:
sum (omni + figure-8)   front

rear channels:
difference (omni - figure-8)   rear

FIGURE 15–10 ——————

The Schoeps KFM 360 four-microphone array: photo of left side of unit showing imbedded pressure microphone and external gradient microphone (A); front and rear pattern selection for the KFM 360 (B). (Photo courtesy of Schoeps.)

The Holophone Global Sound microphone system

This system has a set of five pressure microphone elements located on the surface of an oval-shaped (dual radius) ellipsoid. The microphones can be extended out a short distance from the ellipsoid for enhanced separation. In its wireless form the system is excellent for field effects gathering.

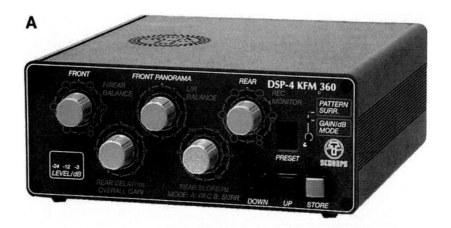

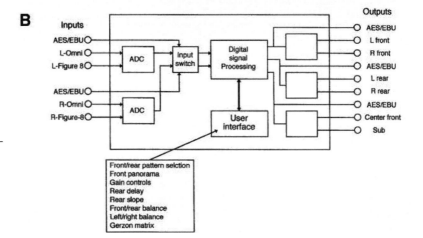

FIGURE 15-11 ———

DSP-4 control unit for
Schoeps KFM 360; photo
of unit (A); signal flow
diagram for control unit
(B). (Data courtesy of
Schoeps.)

SPACED ARRAYS

The surround ambience microphone (SAM) array

An arrangement of four cardioids at 90° has been proposed by Theile of
the German IRT (Institut für Rundfunktechnik). As shown in
Figure 15–12, these microphones are located at the corners of a square
measuring 21 to 25 cm per side (8–10 in). The array is known as SAM
(surround ambience microphone). The spacing of the elements of the
array adds significant time cues to supplement the amplitude cues delin-
eated by the microphone patterns. The array is generally intended to pick
up ambience in live spaces in conjunction with traditional accent microphones center channel pickup (Theile, 1996).

The sound performance lab (SPL) array

Figure 15–13A shows a photo of the German SPL array of five multi-
pattern microphones. The array offers a high degree of flexibility, with

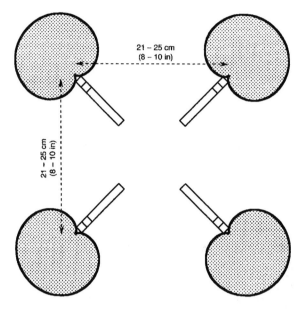

21 – 25 cm
(8 – 10 in)

21 – 25 cm
(8 – 10 in)

FIGURE 15–12 ————

Details of the SAM
cardioid microphone array.

A

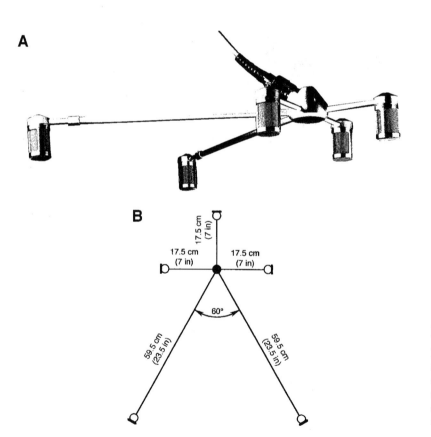

B

17.5 cm
(7 in)

17.5 cm
(7 in)

17.5 cm
(7 in)

60°

59.5 cm
(23.5 in)

59.5 cm
(23.5 in)

FIGURE 15–13 ————

Details of the SPL
microphone array: photo
of array (A); plan view of
array showing normal
dimensions (B). (Photo
courtesy of SPL, USA.)

individual setting of distances between microphones (using telescoping sections) and horizontal aiming directions for each microphone. The orientation shown at B represents normal operating distances among the five microphones. An array such as this may be used to produce a basic spatial "signature" of a recording environment. Accent microphones may be used as needed with the basic array. The control unit provides a number of functions, including microphone pattern switching, ganged level control, sub-bass output control, and various panning functions. Note that the SPL array bears a striking resemblance to the Decca tree, as discussed in Chapter 13.

FRONTAL ARRAYS

In this section we will describe methods of specifically picking up the three frontal signals for optimum effect and separation. These arrays may be used with a variety of pickup methods for general room ambience. One of the problems with three-microphone frontal arrays with each microphone fed to its own frontal loudspeaker is the reproduction of three phantom images produced by the microphones taken two at a time: L-C, C-R and L-C. If there is considerable pattern overlap of the three microphones, the phantom images between the outside pair will conflict with the real image produced by the center loudspeaker. This can be minimized by using hypercardioid or supercardioid patterns for the left and right microphones, which will minimize this tendency

Klepko (1997) describes the microphone array shown in Figure 15–14. Here, the left and right microphones are hypercardioids, and the center microphone is a standard cardioid. The use of hypercardioids for left and right pickup will minimize the strength of any phantom image between the left and right loudspeakers in the playback array.

Schoeps proposes the frontal array shown in Figure 15–15. Known as OCT (Optimum Cardioid Triangle), the array uses increased spacing and supercardioids facing full left and full right to minimize phantom images between outside loudspeakers. The center cardioid element is positioned slightly forward of the two supercardioids to give the center channel signal a slight anticipatory time advantage for frontal pickup, relative to the left and right microphones. The added omni microphones

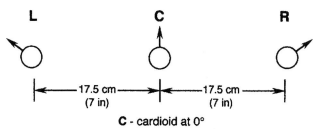

FIGURE 15–14 ——————

Three-microphone frontal array proposed by Klepko.

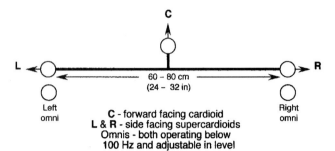

FIGURE 15-15

Three-point frontal array proposed by Schoeps.

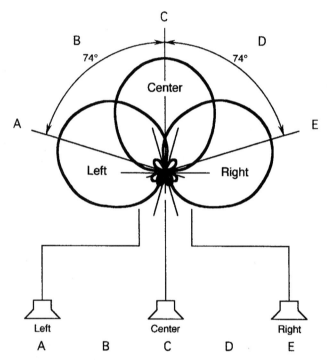

FIGURE 15-16

Details of a coincident array of three second-order cardioids.

at left and right are primarily for restoring the normal LF rolloff caused by the supercardioid microphones.

A second-order cardioid coincident frontal array can be conceived as shown in Figure 15–16 (Cohen and Eargle, 1995). While a number of engineering obstacles still remain in realizing this configuration, the array has pickup characteristics that virtually eliminate phantom images between left and right loudspeakers, while providing uniform pickup over a 150° frontal angle. Adjacent crosstalk along the common left and right microphone axis is about −15 dB, clearly indicating that signal commonality between L and R will be minimum. The equation of the second-order pattern is:

$$\rho = (0.5 + 0.5\cos\theta)(\cos\theta).$$

The foregoing discussion has dealt with minimizing crosstalk, or leakage, from the center channel into the front flanking channels, and the microphone arrays we have shown will accomplish this. However, in many pop mixes there is a preference for a strong frontal presentation of a soloist, and many remix engineers and producers will purposely feed some of the center channel into the left and right flanking channels to accomplish this. The level of this added feed is about 3 dB lower than that present in the center channel, and it produces something resembling a "wall of sound" from the front. This a judgement call on the part of the engineer and producer and is based on musical requirements rather than technical requirements. Proceed with caution when doing this.

CREATING A CENTER CHANNEL FROM A LEFT-RIGHT STEREO PAIR OF CHANNELS

The reader will have noticed that several of the microphone arrays discussed so far have no specific center channel microphone. In these cases, as well as in remixing older stereo program material for surround sound, it is often necessary to synthesize a center channel from a stereo pair. Gerzon (1992) describes a matrix network that accomplishes this, albeit with some loss of overall left-right separation in the frontal three-loudspeaker array. The circuit, shown in Figure 15–17, may look fairly complex, but it can easily be set up on any console that has switchable polarity inversion in its line input sections as well as a flexible patch bay. It can also be set up on the virtual console pages of many computer based editing programs.

Here is an example of its application: Assume we have a stereo recording with three main signal components: left, phantom center and right. The stereo channels may then be represented as:

$$L_T = L + 0.7C$$
$$R_T = R + 0.7C$$

where C represents the panned center signal, such as a vocalist, and L and R represent discrete left and right program elements. L_T and R_T comprise the left-total and right-total stereo program. The matrix separation

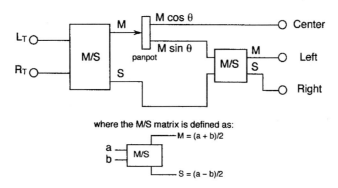

FIGURE 15–17 ———

Gerzon 2-3 matrix for deriving a center channel from a stereo program pair.

angle θ is normally set in the range of 45°, giving values of sin and cos of 0.7. For this value, the three output signals of the matrix will be:

$$\text{Left} = \mathbf{0.85}L + 0.5C - 0.15R$$
$$\text{Center} = 0.5L + \mathbf{0.7}C + 0.5R$$
$$\text{Right} = -0.15L + 0.5C + \mathbf{0.85}R$$

You can see that L, C and R components (shown in boldface) are dominant, respectively, in their corresponding output channels. It is also clear that the crosstalk has increased noticeably. A higher value of θ will increase left–right separation, but at the expense of center channel level. Gerzon recommends that a value of 55° be used at frequencies above about 4 kHz, while a value of 35° be used at lower frequencies. Such an approach actually calls for two matrices and may be needlessly complicated. If you wish to implement the Gerzon matrix, we recommend that you proceed carefully in establishing an acceptable balance.

MICROPHONE ARRAYS FOR SINGLE-POINT PICKUP

Single-point surround pickup arrays are generally intended to map the global sound field in a performance space into a loudspeaker array in the playback space. A coincident stereo microphone pair is, in a manner of speaking, a rudimentary version of this, but generally we think in terms of four or more transmission channels with microphones in a three-dimensional array. The Soundfield microphone, as introduced by Gerzon (1975), uses four microphone to create an array of first-order directional patterns, each of which can be assigned to a loudspeaker position in space matching that of the microphone. Johnston (2000) has suggested an array of seven rifle microphones with a corresponding playback setup. Meyer (2002) proposes a third-order microphone array, known as the Eigenmike, which supports a playback array of 16 loudspeakers.

Questions of practicality are bound to arise when the number of playback channels increases beyond five or six – at least as far as the home environment is concerned. On the other hand, there are many special entertainment venues and occasions where multichannel techniques can be taken to the limit.

THE SOUNDFIELD MICROPHONE

As we discussed in Chapter 5, any first-order cardioid pattern may be produced by combining an omnidirectional pattern with a figure-8 pattern. By selectively combining a single omnidirectional capsule with three figure-8 elements, individually oriented in left-right, up-down, and fore-aft directions, it is possible to produce a first-order cardioid pattern pointed at any direction in space. Gerzon (1975) developed an array of four subcardioid patterns, each oriented parallel to the sides of a regular tetrahedron and contained within the structure shown in Figure 15–18A.

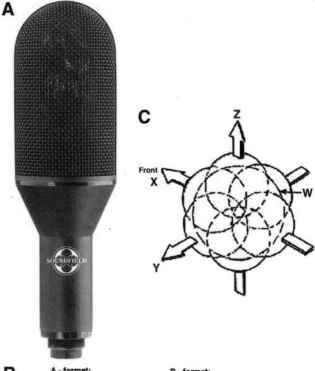

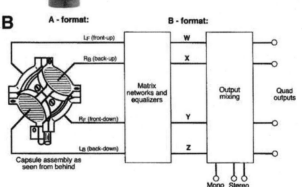

FIGURE 15–18

Details of the Soundfield microphone. Photo of microphone (A); diagram showing signal flow in the control unit (B); details of the B-format (C); photo of control unit (D); detail functions of the control panel (E). (Photos courtesy of Transamerica Audio Group.)

A rear view of the four capsule elements comprising the A-format is shown at B. The array elements occupy a relatively small space, providing balanced resolution in all directions over a wide frequency range. The orientation of the subcardioid elements, as seen from the rear of the assembly, is as follows:

Forward elements: 1. Left-up (L_U)

2. Right-down (R_D)

Back elements: 1. Right-up (R_U)

2. Left-down (L_D)

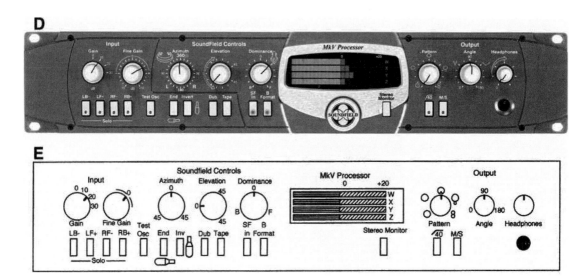

FIGURE 15–18

Continued.

These four A-format elements are combined to produce the four B-format components, as shown:

W = Pressure component = $(L_U + R_D + R_U + L_D)$

X = Fore-aft velocity component = $(L_U + R_D) - (R_U + L_D)$

Y = Left-right velocity component = $(L_U + L_D) - (R_D + R_U)$

Z = Up-down velocity component = $(L_U + R_U) - (R_D + L_D)$

Graphic details of the B-format are shown at Figure 15–18C, and these four elements can be combined to produce the entire range of the first-order cardioid family oriented in any direction. A photo of the front panel of the control unit is shown at D, and further details of the individual controls are shown at E. Normally, four separately resolved output are used for surround sound applications, along with mono and stereo outputs.

In its stereo (two-channel) mode of operation, the following manipulations can be carried out remotely at the control unit:

1. Rotation of the patterns, with no physical contact or manipulation of the microphone assembly itself.

2. Electrical adjustment of the stereo pickup plane for any desired downward tilt angle, with no physical movement involved.

3. A front-back dominance control to "focus" the stereo pickup toward the front (for greater soloist dominance) or toward the back (for greater reverberant pickup).

As such, the Soundfield microphone has great application as a permanent fixture in auditoriums and concert venues in schools of music and at music festivals, where many instrumental groups, each requiring its own stereo microphone pickup setting, may be used during a single program with virtually no interruption in the flow of things. To facilitate stereo applications, the Soundfield microphone may be used with a simplified stereo-only control unit.

When used in connection with surround sound presentation, the Soundfield microphone's four primary outputs can be positioned anywhere in the listening space and B-format settings adjusted so that the output from each of the four (or more) loudspeakers corresponds exactly to that of a first-order microphone oriented in that direction.

THE JOHNSTON-LAM SEVEN CHANNEL ARRAY

Figure 15–19 shows a perspective view of the Johnston-Lam (2000) microphone array. Five hypercardioid microphones are oriented at 72° intervals in the horizontal plane. The vertical microphones are short line models with broad response nulls at 90° to minimize the amount of lateral signal pickup in those channels. All seven microphones are located on a sphere with a diameter of approximately 290 mm (11.5 in). In normal practice the array is placed at a typical listening position in a performance space, elevated about 3 m (10 ft) above the floor.

According to the authors, microphone spacing, orientation, and selection of response patterns are chosen to preserve the interaural time differences (ITD) and interaural level differences (ILD) picked up by the microphones and which are subsequently heard by a listener located at or near the center of the playback loudspeaker array. The ITD and ILD

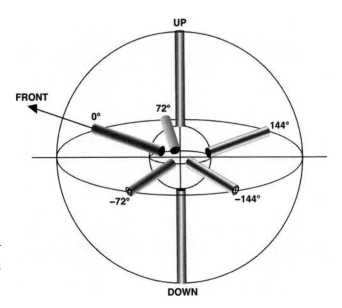

FIGURE 15–19 ———

A view of the Johnston-Lam multichannel microphone array.

are the primary cues the ears need for localization, and the authors suggest that recreating them in this manner is a more realistic goal than attempting to recreate the minute details of the actual sound field itself. The authors state that the approach is essentially compatible with playback over a standard ISO surround loudspeaker array if the height information is appropriately mixed into the playback array. Ideally, the loudspeakers should be positioned at 72° intervals about the listener.

OVERVIEW OF THE EIGENMIKE™

Meyer and Agnello (2003) describe a relatively small spherical array about 75 mm (3 in) in diameter that contains 24 miniature omnidirectional electret microphones equally spaced on its surface. These microphone outputs can be selectively combined and adjusted in relative delay to create first, second, or third-order pickup patterns arrayed equally in three dimensions.

The required number of microphones for optimum spherical coverage is given by:

$$\text{Number of microphones} = (N + 1)^2 \qquad (15.1)$$

where N is microphone order. For example, in the Soundfield microphone first-order array the equation gives a value of 4, and as we have seen there four elements in the B-format from which all remaining patterns are derived. In a second-order system there are 9 elements, and in a third-order there are 16 elements. Figure 15–20 shows a view of the Eigenmike.

FIGURE 15-20 ———

Photograph of the Eigenmike. (Photo courtesy of mh-acoustics, Summit, NJ.)

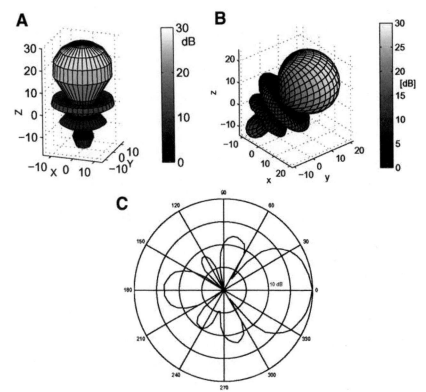

FIGURE 15-21

Formation of Eigenbeams: third-order pattern along one axis (A); third-order pattern at θ and φ each of 20° (B); a two-dimensional polar plot of a third-order hypercardioid pattern at 3 kHz (C). (Data courtesy of Meyer and Agnello, 2003.)

In a manner analogous to the *A* and *B* formats that used in the Soundfield microphone, the outputs of the multiple elements are combined in two stages. The resulting flexibility here allows *Eigenbeams*, or directional microphone elements, to be formed continuously over the sphere independently of the number of actual sensors. Figure 15–21A shows a synthesized third-order hypercardioid pattern in three dimensions and positioned along one axis. At B is shown the same pattern positioned at θ and φ angles, each of 20°. A two-dimensional view of the pattern for 3 kHz is shown at C.

In a typical application involving an array intended for sound field reconstruction, third-order patterns would maintained above about 1.5 kHz. At progressively lower frequencies the directivity would be reduced to second-order and finally, below 700 Hz, to first-order. The compromise here is between spatial resolution and LF noise, as discussed in Chapter 6. Alternatively, the entire array size can be increased to offer a better set of tradeoffs between pattern control and noise at LF.

In addition to audio applications the array has great promise in acoustical architectural measurement applications, where multichannel recording can be used to store multiple impulse data for later directional analysis off-line.

TRANSAURAL (HEAD RELATED) TECHNIQUES

Cooper and Bauck (1989) use the term *transaural* to describe sound recording and reproduction systems in which the directional response is determined by conditions existing independently at the listener's ears. By comparison, normal stereo reproduction is defined principally in terms of what comes out of the loudspeakers, and of course the sound from each loudspeaker is heard by both ears.

Transaural recording and playback is a direct outgrowth of binaural sound, which we discussed in Chapter 12. The basic plan is shown in Figure 15–22, where the two channels of sound picked up by an artificial head are fed through a crosstalk canceller. In this stage, delayed signals from each loudspeaker are cross-fed and arrive at opposite ears with the exact delay time required to cancel the normal stereo leakage, or crosstalk, signals.

The details of the Schroeder-Atal crosstalk cancelling circuit are shown in Figure 15–23, where the binaural signal pair is transformed by the two signals labeled C, where $C = -A/S$. S and A are, respectively, the transfer functions from the loudspeakers to listener for the near-side signal (S) and the far-side signal (A) at the ears. The performance of the crosstalk-cancelling circuitry is normally set for a given listening angle (θ), and the system's cancellation performance will be optimum for

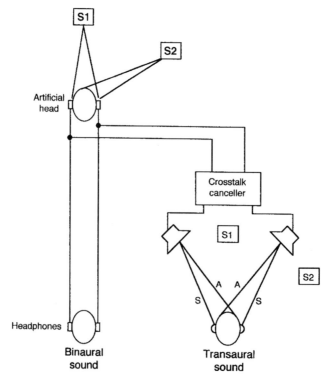

FIGURE 15–22

Transaural technology: a binaural signal can be played back directly via headphones or over a pair of loudspeakers using crosstalk cancellation.

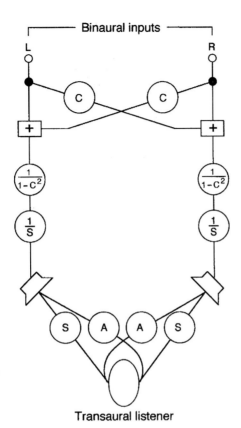

Binaural inputs

FIGURE 15-23

Details of the
Schroeder-Atal crosstalk
cancellation network.

loudspeakers positioned at that angle. Off-axis listening, or listening on-axis at other angles, will result in non-optimum performance.

Transaural processing works most convincingly when the A and S data are actually the same as those measured for a given listener – that is, when the listener's head is actually substituted for the artificial head in making the measurements. Generally, however, measurements made on a typical artificial head will give results that are a good approximation to the shape of an average adult head. In some laboratory playback systems, individual head-related transfer functions (HTRFs) can actually be accessed for improved performance.

For studio use Cooper and Bauck have developed the panning system shown in Figure 15–24. Using impulse response methods, the HTRFs for a given head are measured at each desired azimuthal position about the head (A). These measurements are stored and can be introduced into a panning system (B) that will translate a given input signal to a specific position over the frontal 180° listening arc as perceived by a listener. Positioning can be synthesized for each 5° angular increment.

Using a system such as this, a two-channel transaural presentation can be generated from a five-channel surround program in which the listener will hear each of the five channels originating from its proper

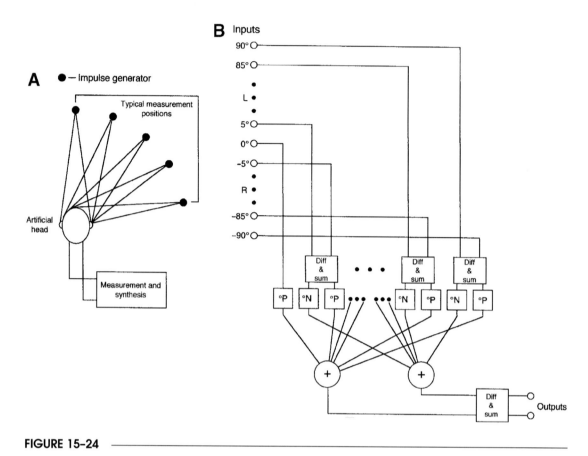

FIGURE 15-24

Details of a transaural panning scheme: making HTRF measurements (A); overall detail of the panner (B). (Data after Cooper and Bauck, 1989.)

position in virtual space. The effect we have described is as shown in Figure 15-25.

A note on performance: transaural virtual images are only as stable as the listener's head is stationary. A pair of loudspeakers is required for each listener, and if listeners move their heads from side to side, the virtual images will move with them. With the aid of a technique called *head tracking*, compensating signals can be fed back to the synthesizer, partially correcting the tendency for the images to move when the listener's head rotates from side to side.

The transaural technique has great promise when used in conjunction with computer monitors and television sets. Today, many TV programs are broadcast in surround sound that has been encoded for transaural playback. For TV loudspeakers located about 0.5 m (20 in) apart, a listening distance of approximately 2 m (6.7 ft) will usually suffice to give the listener a good impression of what transaural techniques can provide. An on-axis location is critical, and normally only a single listener can

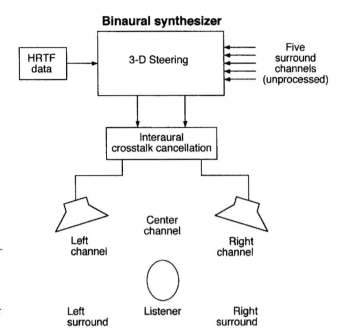

Binaural synthesizer

HRTF data → 3-D Steering → Five surround channels (unprocessed)

Interaural crosstalk cancellation

Left channel   Center channel   Right channel

Left surround   Listener   Right surround

FIGURE 15–25

Overall view of a transaural presentation of a five-channel surround program using a stereo pair of channels.

appreciate the effect to the fullest extent. The same technology is often applied to video games and other computer programs. Here, the viewing distance is much closer than is the norm for TV, and appropriate adjustments will have to be made.

Transaural systems have generally required extensive internal signal processing, but the costs of such operations continues to drop, making them more accessible to modern application in listening environments, such as computer workstations, where listening positions are stable and can be accurately estimated.

SYSTEMS WITH PARALLAX

In a landmark paper on motion picture sound technology, Snow (1953) described the "perfect" record/playback system shown in Figure 15–26. A horizontal array of microphones communicate, each independently, to a horizontal array of loudspeakers in the theater environment. A diverging wavefront at the microphones creates a sequentially firing set of individual wavefronts from the loudspeakers which coalesce, via Huygen's law on wavefront reconstruction, into a near copy of the original wavefront impinging on the microphones. We may think of this as an acoustical hologram in which listeners in the theater are free to move around and hear individual acoustical sources in their proper left-right, fore-aft relationships – in other words, as fixed in space as real sources.

The hardware requirements are staggering to consider, and such a system has long been little more than a dream. However, modern technology has come to the rescue with multiple small loudspeakers, small

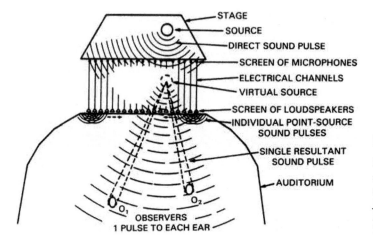

FIGURE 15-26

Snow's multichannel solution to motion picture stereo. (Data courtesy of *Journal of the Society of Motion Picture and Television Engineer*.)

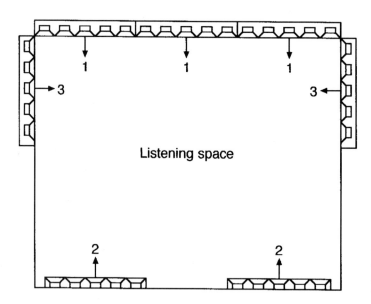

FIGURE 15-27

Playback environment for holographic sound presentation. (Data after Horbach, 2000.)

amplifiers, and relatively low-cost digital signal processing capability. Horbach (2000) describes one approach to this.

A playback environment is visualized as shown in Figure 15–27. Here, a "wraparound" loudspeaker array is placed in front and smaller linear arrays are placed along the back wall. Ideally, we would like the loudspeakers to be small and very closely packed; in practice, one loudspeaker every 15 cm (6 in) is about as dense as we can reasonably get.

The data that is actually recorded consists only of the relative dry, unreverberated instrumental tracks and other basic musical elements produced acoustically in the studio. Details of the overall acoustics of the recording space will have been gathered using impulse measurements, and this added data, along with the actual music tracks, will be used to reconstruct the recording during the playback process.

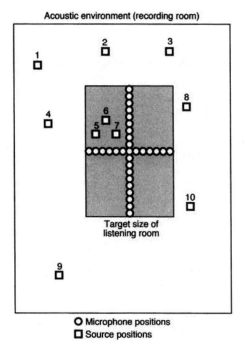

Acoustic environment (recording room)

Target size of
listening room

○ Microphone positions
□ Source positions

FIGURE 15-28

Recording environment for holographic sound pickup. (Data after Horbach, 2000.)

A typical recording setup is shown in Figure 15–28. Here, a large number of omnidirectional microphones are arranged in crossed line arrays as shown. The small squares indicate the target locations of individual instruments in the recording space.

The first step is to build up a large library of impulse responses. At this stage, impulse signals are produced at each instrument location (indicated by the numbered squares in the figure). For each studio location, impulse measurements will be made at *all* of the microphone locations. If there are 10 studio locations and 31 microphone locations, there will be 10×31, or 310 impulse response files. Each impulse file is relatively small and of course needs to be made only once. Other impulse files detailing the overall studio reverberation characteristics may be made as well.

When the final program mix is constructed, the engineer and producer bring together all of the individual recorded source tracks and all of the impulse data files. A stage plan is then laid out with each instrument or musical element assigned a specific position, corresponding to one of the numbered squares in Figure 15–28. The impulse files that define the sound stage boundaries for that signal position are then accessed and assigned to a specific playback channel. When all of the impulse files are convolved with the audio track and monitored over the multi-loudspeaker playback array, the producer and engineer will hear the audio track positioned accordingly behind the loudspeaker array. This assignment process is repeated until all of the audio tracks have been appropriately positioned on the virtual stage.

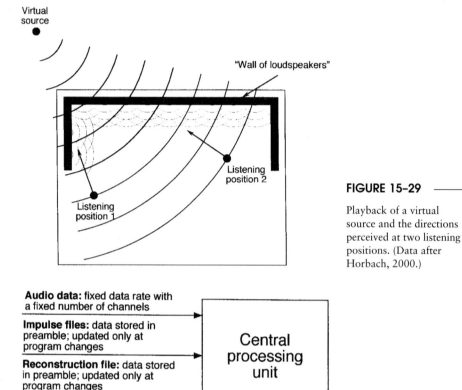

FIGURE 15–29

Playback of a virtual source and the directions perceived at two listening positions. (Data after Horbach, 2000.)

FIGURE 15–30

Overview of signal flow in a system with parallax.

Figure 15–29 shows how a typical virtual source is made to appear fixed in a given location behind the loudspeaker array. Within certain limitations, it is also possible to position virtual sources in front of the loudspeaker array.

GENERAL COMMENTS

The methods we have just described are more akin to modern pop studio music production techniques than to verbatim recreation of natural acoustical sound fields. While the basic data transmission rate from studio to playback is not inordinately high (and carries with it many opportunities for beneficial data reduction), the actual on-site playback data processing rate can be quite high. As shown in Figure 15–30, only the raw audio information is continuous; impulse data files and reconstruction information are one-time only events and can be downloaded via a preamble to the program file. Needless to say, the playback hardware must be efficiently designed to handle all of this on a timely basis.

C H A P T E R   1 6

SURROUND RECORDING CASE STUDIES

INTRODUCTION

In this section we will examine six surround recordings that demonstrate a number of the techniques discussed in the previous chapter. These examples range from fairly modest studio resources up to large concerted works that include chorus, organ, piano, and percussion ensembles in performance with normal orchestral resources. In each case, recording channel limitations restricted us to no more than eight tracks. This called for careful planning, since the eight "stems" had to be used not only for creating surround mixes but also for making any adjustments in the stereo mix during later postproduction. In this connection we will introduce a concept of *subtractive* mixing, as it applies to *demixing* and subsequent *remixing*, of tracks to attain an ideal balance.

All of the projects studied here were targeted for both stereo and standard 5.1 channel playback. In each case we will present the basic microphone choice and layout, channel assignments, mixing assignments in both stereo and surround, and an overall commentary. All of the projects discussed here have been commercially released, and a discography is included at the end of this chapter.

The items to be discussed are:

1. Berlioz: March to the Scaffold from *Symphonie Fantastique* (orchestra)
2. Tchaikowsky: *1812 Overture* (orchestra with chorus)
3. Gershwin: *Rhapsody in Blue* (orchestra and piano)
4. Berlioz: *Te Deum* (orchestra with chorus, vocal soloist, and organ)
5. Bizet-Shchedrin: *Carmen Ballet* (strings and large percussion ensemble)
6. Schnittke: *Piano Concerto* (strings and piano)

A SURROUND MIXING RATIONALE: DEFINING THE SOUND STAGE

What is the purpose of a surround sound musical presentation? One answer, especially for classical music, is to present that music in the context of the original performance venue with its acoustics intact. For pop and rock music another goal may be to place the listener in the center of an imaginary stage, with sound elements widely positioned all around. Both are respectable goals, and there should be no prejudice favoring one over the other.

The recordings described here all fit neatly into the former, or so-called direct-ambient, approach in which there is a stage at the front and an enveloping ambience, based on the peformance space, that reaches the listener from all directions. This does not happen casually, and there are some general procedures that should be followed. They are:

1. *Controlled monitoring environment:* An ITU standard loudspeaker monitoring array was used with reference levels set to within 0.25 dB. Even though the mix was made with engineer and producer sitting at the "sweet spot," the program was auditioned at various points within the periphery of the loudspeakers to ensure that it was generally and broadly effective.

2. *Decorrelated ambience:* The basic ambient signature of the performance space is presented over all loudspeakers except the front center, and these four ambience signals are all decorrelated, either through signal processing or through the original spacing of microphones.

3. *Control of early reflections:* In any performance venue, early reflections arising primarily from side walls help to define the boundaries of the space and convey a sense of room size. If the reflections are too early and too high in level, they may add muddiness and confusion. The optimum delay range for them is between 25 and 40 milliseconds, presented at a level between −10 and −15 dB relative to the primary signal. If we're lucky, these signal components are produced directly in the room and are picked up naturally; at other times, the engineer has to create them via signal processing using modern delay and reverberation generators. These signals are normally introduced into the mix only in the left and right-front channels as well as the two rear channels. In some cases it is effective to delay the rear signals slightly, relative to the front left and right signals. You will see application of these techniques as we progress through this chapter.

4. *Layers:* When multiple microphones have been used in a recording – and this is the norm today – it is often helpful to think in terms of structural *layers* in the recording. For example, the main frontal microphones (and soloist microphones if appropriate) comprise a single layer. The ensemble of accent, or spot, microphones in the

orchestra will comprise another layer, and the house microphones, which convey only reverberation and ambience, will comprise yet another layer. Normally, if level changes are made in postproduction, the entire layer will be altered as a unit in order to avoid any skewing of the sound stage. During the recording process however individual microphones, or microphone pairs, may be altered on a running basis as dictated by musical requirements.

5. *Track assignments and allocation:* When using fewer recording tracks than there are microphones, informed decisions must be made regarding which microphones to combine when laying down the initial tracks. In general the decision is made to subdivide the recorded elements into basic pre-mixed tracks, or *stems*, which can be more easily manipulated during postproduction. Since all of the works recorded here were to be released first in stereo, a decision was made to aim for a finished live-to-stereo mix on tracks 1 and 2, while ensuring that the remaining stems yielded enough flexibility for creating surround mixes as well as for making minor adjustments in the stereo mix itself. While all eight tracks were edited at the same time, the stereo tracks were used as as the basic guide for all edit decisions.

"MARCH TO THE SCAFFOLD," FROM *SYMPHONIE FANTASTIQUE* BY HECTOR BERLIOZ

The recording venue was the New Jersey Performing Arts Center, a 2500-seat hall with a midrange reverberation time of about 2.3 s. Zdenek Macal conducted the New Jersey Symphony Orchestra. Diagrams of the stage layout are given in Figure 16–1, showing positions of the microphones. The recording was made over three evenings with audience present and was to be used for both stereo CD production and release in surround sound. The direct-to-stereo mix was monitored at the sessions and fed directly to channels 1 and 2 of the six-channel recorder (20-bit/48 k sampling). Details of microphone deployment are shown in Table 16–1.

Note the use of only two omnidirectional microphones in the setup. This is in keeping with the principle stated in the chapter on classical recording in which the main ORTF pair and its associated flanking omnis provide the basis of the pickup. All other microphones are secondary to the main four and are intended for subtle heightening of instrumental and sectional presence and balance. As such, these microphones function best with the side and back rejection that the cardioid pattern offers.

The accent microphones were all fed, in varying amounts, to a digital reverberation side chain whose parameters were adjusted to match those of the hall itself. The stereo returns from the reverberation generator were panned left and right into the overall stereo mix. The house microphones were aimed at the upper back corners of the hall to minimize pickup of direct sound from the stage.

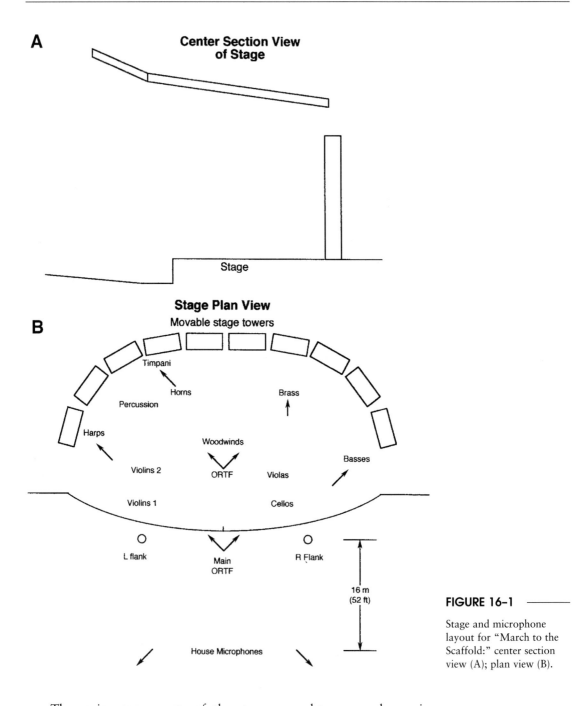

FIGURE 16-1

Stage and microphone
layout for "March to the
Scaffold:" center section
view (A); plan view (B).

The major components of the stereo soundstage are shown in
Figure 16–2A. In creating the surround soundstage, it was necessary to
"demix" the main ORTF pair from the stereo array and re-pan them into
the frontal sound loudspeaker array as shown in Figure 16–2B. This was
done using a technique known as *subtractive mixing*, which can be

explained as follows:

Let the main array consist of the following four components:

$$L_T = L_F + gL_M$$
$$R_T = R_F + gR_M,$$

where L_F and R_F are the outputs of the left and right flanking microphones, and L_M and R_M are the outputs of the left and right main ORTF

TABLE 16-1 Microphone Deployment for "March to the Scaffold"

Position	Description	Stereo panning	Mic height	Track assignment
Major components				
Stereo mix left				Track 1
Stereo mix right				Track 2
L flank	omni	left	3.5 m (12 ft)	
L ORTF	cardioid	left	3.5 m (12 ft)	Track 3
R ORTF	cardioid	right	3.5 m (12 ft)	Track 4
R flank	omni	right	3.5 m (12 ft)	
L house	cardioid	left	4 m (13.3 ft)	Track 5
R house	cardioid	right	4 m (13.3 ft)	Track 6
Accent microphones (these appear only in the stereo mix)				
L woodwinds	cardioid	half-left	3.5 m (12 ft)	
R woodwinds	cardioid	half-right	3.5 m (12 ft)	
Harps	cardioid	left	1 m (40 in)	
Timpani	cardioid	half-left	2 m (80 in)	
Brass	cardioid	half-right	3 m (10 ft)	
Basses (1st stand)	cardioid	right	2 m (80 in)	

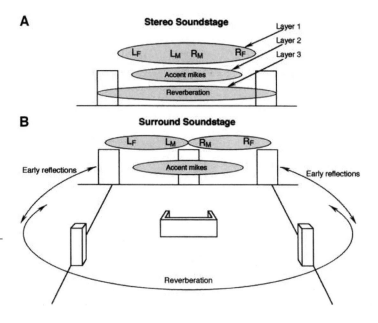

FIGURE 16-2

Target recorded soundstages for "March to the Scaffold:" stereo (A); surround sound (B).

microphones. The symbol g represents a gain factor for the two ORTF signals, and it is carried over into the signals recorded on tracks 3 and 4 of the digital recorder.

The next step is to *subtract* the ORTF pair from L_T and R_T. This can be done because the value of g is the same in tracks 1 and 2 as it is in tracks 3 and 4; the result is:

$$L_T = L_F + gL_M - gL_M = L_F$$
$$R_T = R_F + gR_M - gR_M = R_F$$

Having "stripped out" the ORTF pair from the left and right frontal channels of the surround array, we are free to reposition those signals anywhere in the array we wish. The usual approach is to reintroduce them by panning L_M slightly left of the center channel and R_M slightly right of the center channel, as shown in Figure 16–2B. As a matter of practicality, it is possible to carry out subtractive mixing on any console that has phase reversal stages in each input strip and the facility for patching and multing various signals. With practice, the process can be easily carried out by ear, listening for the "null zone" as you alternately raise and lower the level of the antiphase components.

A functional view of the subtractive mixing process is shown in a basic form in Figure 16–3. Here, four tracks have been recorded: a stereo mix (L_T and R_T) on tracks 1 and 2, along with the main ORTF

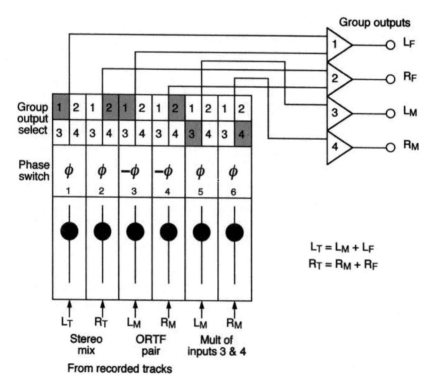

$$L_T = L_M + L_F$$
$$R_T = R_M + R_F$$

FIGURE 16-3

Functional view of the subtractive mixing process.

pair (L_M and R_M). Through subtractive mixing we want to separate the flanking microphone signals (L_F and R_F) from the stereo mix. This is accomplished by feeding the ORTF pair into faders three and four in antiphase. When these are then fed into Groups 1 and 2, the L_M and R_M signals will be canceled, leaving only L_F and R_F at the outputs of Groups 1 and 2. For the cancellations to be complete the levels must be matched. This may be done by adding test signals at the head of all recorded tracks before the session begins, ensuring that all gain settings downstream from the faders are fixed. With a little practice, you can approximate this condition by raising and lowering the phase-inverted faders through the null region and then isolating the actual null position by ear.

Having adjusted the frontal array as discussed above, we then move on to the requirements of the rear channels. In this recording, the distance of the house microphones from the main array resulted in a relative delay of about 44 ms, which is too great by about 20 ms. This value was compensated for on the virtual console of the digital editing system by delaying *all signals* other than the house microphones, producing a net time onset of reverberation of 24 ms that was within normal bounds for good surround performance. The final step was to re-introduce signals from the main ORTF pair into the back channels to simulate added early reflections from the sides. To do this, the outputs of L_M and R_M were high-passed at about 200 Hz, delayed about 20 ms, reduced in level and fed respectively to the left-rear and right-rear channels.

1812 OVERTURE BY PYOTR TCHAIKOWSKY (CHORAL ARRANGEMENT BY IGOR BUKETOFF)

The recording venue was McDermott Hall in the Meyerson Symphony Center, Dallas, TX. Andrew Litton conducted the Dallas Symphony Orchestra and Chorus. As before, the recording was intended for both stereo and surround sound release. The recorded track assignment was composed of: stereo mix (tracks 1 and 2), main pair (tracks 3 and 4), flanking pair (tracks 5 and 6) and house pair (tracks 7 and 8). Figure 16–4 shows stage and microphone layouts. The composite stereo monitor mix was assigned to channels 1 and 2 of the digital recorder. Details of microphone deployment are shown in Table 16–2.

The basic pickup is very much like that of the previous example. The use of omnidirectional microphones for chorus pickup is a judgemental choice; the broad omni pattern ensures that more members of the ensemble will be covered than if cardioids were used. A drawback here is the increased leakage of brass and percussion (back of the orchestra) into the chorus microphones, a problem that required carefully planned gain adjustments during the sessions.

In the *a capella* choral opening of the work, the flanking microphones were panned to the rear loudspeakers, giving the impression of a chorus placed at the back of the hall. As the introduction progressed,

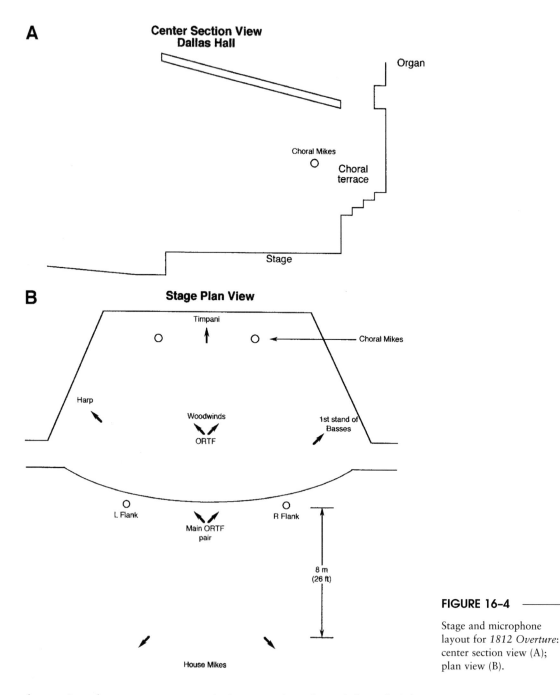

A

**Center Section View
Dallas Hall**

Organ

Choral Mikes

Choral
terrace

Stage

B   **Stage Plan View**

Timpani

Choral Mikes

Harp

Woodwinds

1st stand of
Basses

ORTF

L Flank   R Flank

Main ORTF
pair

8 m
(26 ft)

House Mikes

FIGURE 16-4

Stage and microphone
layout for *1812 Overture*:
center section view (A);
plan view (B).

these microphones were progressively panned to front left and right.
Stage details are shown in Figure 16–4. The stereo and surround soundstages are shown in Figure 16–5. In setting up the surround mix, the
main microphone pair was demixed and re-panned center left and right
in order to produce a center channel. All accent microphones appear

TABLE 16-2   Microphone Deployment for *1812 Overture*

Position	Description	Stereo panning	Stage height	Track assignment
Major components				
Stereo mix left				Track 1
Stereo mix right				Track 2
L flank	omni	left	3.5 m (12 ft)	Track 7
L ORTF	cardioid	left	3.5 m (12 ft)	Track 3
R ORTF	cardioid	right	3.5 m (12 ft)	Track 4
R flank	omni	right	3.5 m (12 ft)	Track 8
L chorus	omni	left	4.5 m (14.5 ft)	
R chorus	omni	right	4.5 m (14.5 ft)	
L house	cardioid	left	4 m (13.3 ft)	Track 5
R house	cardioid	right	4 m (13.3 ft)	Track 6
Accent microphones (these appear only in the stereo mix)				
L woodwinds	cardioid	half-left	3.5 m (12 ft)	
R woodwinds	cardioid	half-right	3.5 m (12 ft)	
Harp	cardioid	left	1 m (40 in)	
Timpani	cardioid	half-left	2 m (80 in)	
Basses (1st stand)	cardioid	right	2 m (40 in)	

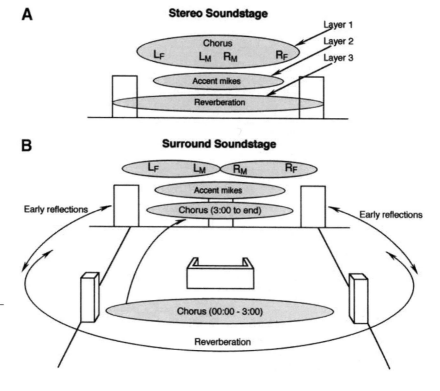

FIGURE 16-5

Recorded soundstages for *1812 Overture*: stereo soundstage (A); surround soundstage (B).

only in the front left and right channels. The house microphones were located 8 m (27 ft) from the main pair and were panned into the rear channels. They were aimed at the upper back corners of the house.

The opening of this work as auditioned in surround sound is an excellent example of the condition shown in Figure 15–3, in which all channels contain mutually uncorrelated reproduction of the chorus. Positioning the chorus at the back of the surround soundstage during the open measures of the work was purely experimental – but is musically appropriate and does not sound contrived.

RHAPSODY IN BLUE BY GEORGE GERSHWIN

The work was performed by Andrew Litton, pianist and conductor, and the Dallas Symphony Orchestra in McDermott Hall at the Meyerson Symphony Center. The normal orchestral forces were reduced to 25 players, in keeping with the original arrangement of the work for the Paul Whiteman Orchestra in 1924. In this regard the recording is more like that of a studio orchestra than a full symphonic ensemble, and microphone distances were reduced slightly for more presence. The cover of

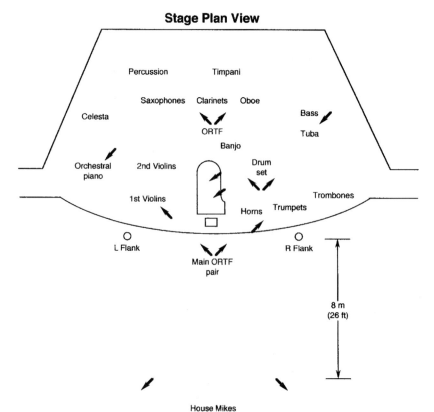

FIGURE 16–6

Stage and microphone layout for *Rhapsody in Blue*.

the piano was removed and the instrument oriented as shown so that the conductor–pianist had direct eye contact with the players. Close pickup of the piano, within 0.5 m (20 in), was mandated by the high ambient orchestral levels surrounding the instrument. Figure 16–6 shows the stage and microphone layout. Microphone deployment details are shown in Table 16–3. The stereo monitor mix was assigned to tracks 1 and 2 of the digital recorder.

Details of the stereo and surround soundstages are shown in Figure 16–7.

The solo piano is a dominant feature in this recording and as such needs to be positioned slightly in front of the orchestra. After some experimenting in postproduction it was felt that the direct sound of the instrument needed to be present in all three frontal loudspeakers, as discussed in Chapter 15 under Frontal Arrays. By using only partial subtractive mixing, an approximate balance was reached, as shown in Table 16–4.

The relationships given here produce a total relative level of 0 dB, and the 3 dB advantage of the center channel ensures that the listener will identify the source primarily at the center loudspeaker. The rear channels were derived primarily from the two house microphones, supplemented by a reduced signal, high-passed at 200 Hz, from the main left and right ORTF pair.

TABLE 16–3 Microphone Deployment for *Rhapsody in Blue*

Position	Description	Stereo panning	Stage height	Track assignment
Major components				
Stereo mix left				Track 1
Stereo mix right				Track 2
L flank	omni	left	3 m (10 ft)	
L ORTF	cardioid	left	3 m (10 ft)	Track 3
R ORTF	cardioid	right	3 m (10 ft)	Track 4
R flank	omni	right	3 m (10 ft)	
L solo piano	cardioid	left	0.5 m (20 in)	Track 7
R solo piano	cardioid	right	0.5 m (20 in)	Track 8
L house	cardioid	left	10 m (33 ft)	Track 5
R house	cardioid	right	10 m (33 ft)	Track 6
Accent microphones (these appear only in the stereo mix)				
L woodwinds	cardioid	half-left	3 m (10 ft)	
R woodwinds	cardioid	half-right	3 m (10 ft)	
Orchestral piano	cardioid	left	1 m (40 in)	
Violins	cardioid	half-left	2 m (80 in)	
Brass	cardioid	half-right	2 m (80 in)	
Bass and tuba	cardioid	right	2 m (80 in)	
L drum set	cardioid	half-left	1 m (40 in)	
R drum set	cardioid	half-right	1 m (40 in)	

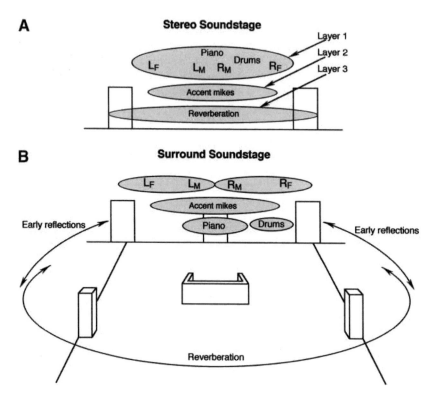

FIGURE 16-7

Recorded soundstages for
Rhapsody in Blue: stereo
soundstage (A); surround
soundstage (B).

TABLE 16-4 Solo Piano Balance in *Rhapsody in Blue*

	Front Left	Front Center	Front Right
Piano level	−6 dB (1/4 power)	−3 dB (1/2 power)	−6 dB (1/4 power)

OPENING HYMN FROM *TE DEUM* BY HECTOR BERLIOZ

Berlioz's *Te Deum* is a large work for orchestra, chorus, tenor solo, and organ. It was performed with audience at the Cathedral of Saint John the Divine in New York in 1996 during a convention of the American Guild of Organists, Dennis Keene, conductor. The Cathedral is the largest gothic structure in the world, with a nave extending 183 m (601 ft) front to back, and the reverberation time of the occupied space is about 5 s. The orchestra was positioned in the large crossing, and the chorus was located on multiple risers in the sanctuary. The vast dimensions of the space are such that there are virtually no early reflections. The first reflections are from walls that are about 25 m (83 ft) distant, so that the reflected sound blends in comletely with the onset of diffuse reverberation.

As the postproduction of the recording got underway it was apparent that the chorus sounded "very present and very distant" at the same time. The obvious solution was to add ambience to the recording using a program in the Lexicon model 300 reverberation generator. This program adds only a set of simulated early reflections and as such gave an effect of immediacy and naturalness not present in the basic recording.

Another problem we faced was where to put the house microphones. The audience in the nave numbered 4000, and there was no way to position stereo microphones 40 m (130 ft) above them. The solution here was to position the house microphones 25 m (80 ft) above the chorus adjacent to the organ pipes. As a result of this, you will hear the organ

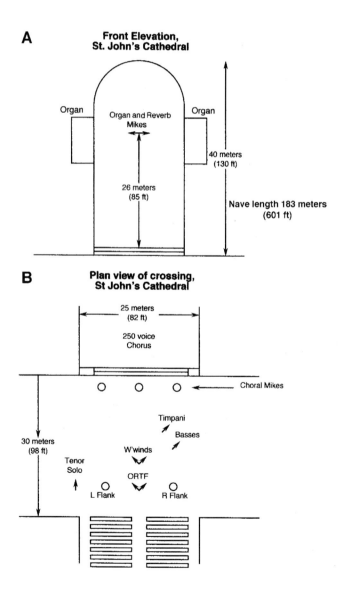

FIGURE 16–8

Stage and microphone layout for *Te Deum*: elevation view (A); plan view (B).

TABLE 16–5   Microphone Deployment for *Te Deum*

Position	Description	Stereo panning	Stage height	Track assignment
Major components				
Stereo mix left				Track 1
Stereo mix right				Track 2
L flank	omni	left	3 m (10 ft)	
L ORTF	cardioid	left	3 m (10 ft)	
R ORTF	cardioid	right	3 m (10 ft)	
R flank	omni	right	3 m (10 ft)	
L chorus	omni	left	3.5 m (11.5 ft)	Track 3
C chorus	omni	center	3.5 m (11.5 ft)	Tracks 3/4
R chorus	omni	right	3.5 m (11.5 ft)	Track 4
L house and organ	cardioid	left	25 m (83 ft)	Track 5
R house and organ	cardioid	right	25 m (83 ft)	Track 6
Vocal solo	cardioid	center	2 m (80 in)	Tracks 7/8
Accent microphones (these appear only in the stereo mix)				
L woodwinds	cardioid	half-left	3 m (10 ft)	
R woodwinds	cardioid	half-right	3 m (10 ft)	
Basses	cardioid	right	2 m (80 in)	
Timpani	cardioid	center	1.5 m (5 ft)	

and reverberant signature of the Cathedral both from the rear channels, and this is very much the way the work was originally heard in Paris in 1855, where the organ was in the rear gallery.

As with the previous examples, the stereo monitor mix was recorded on tracks 1 and 2 of the digital recorder. Figure 16–8 shows views of the recording and microphone setup, and Table 16–5 shows the deployment of microphones. Figure 16–9 shows the resulting stereo and surround sound stages.

SELECTIONS FROM *CARMEN BALLET* BY BIZET-SHCHEDRIN

The Carmen Ballet is an arrangement of music from the opera *Carmen* for strings and large percussion resources. The recording was made by the Monte Carlo Philharmonic Orchestra, conducted by James DePreist, 24–27 June 1996, in the Salle Garnier in Monte Carlo, Monaco. Figure 16–10 shows views of stage and microphone layout, and Table 16–6 shows microphone and track deployment.

The engineer and producer made the decision early on to assign each of the percussion microphones to a separate track. This provided utmost flexibility in rebalancing any of the percussion resources as they might crop up later in postproduction. In a sense, the percussion elements were

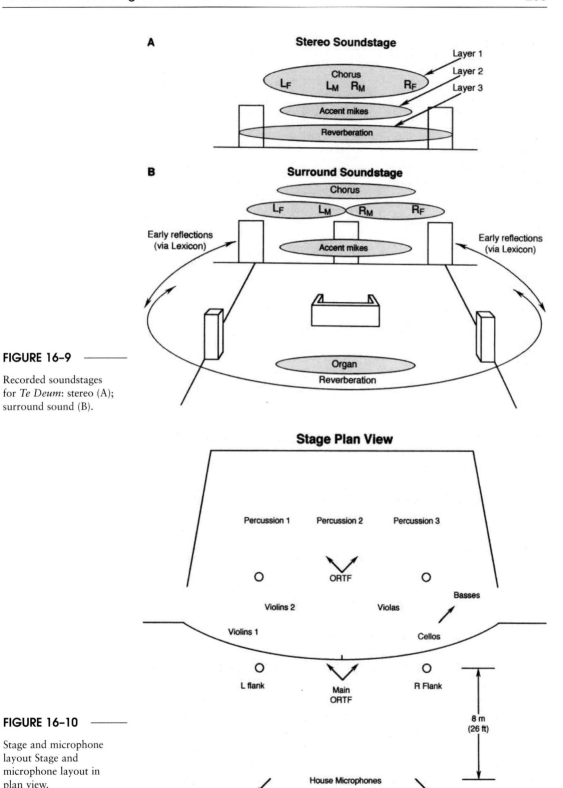

A

Stereo Soundstage

Layer 1
Layer 2
Layer 3

Chorus
L_F L_M R_M R_F

Accent mikes

Reverberation

B

Surround Soundstage

Chorus

L_F L_M R_M R_F

Early reflections
(via Lexicon)

Early reflections
(via Lexicon)

Accent mikes

Organ
Reverberation

FIGURE 16-9

Recorded soundstages for *Te Deum*: stereo (A); surround sound (B).

Stage Plan View

Percussion 1   Percussion 2   Percussion 3

ORTF

Violins 2   Violas   Basses

Violins 1   Cellos

L flank   Main   R Flank
                 ORTF

8 m
(26 ft)

FIGURE 16-10

Stage and microphone layout Stage and microphone layout in plan view.

House Microphones

TABLE 16-6 Microphone and track deployment for *Carmen Ballet*

Position	Description	Stereo panning	Mic height	Track assignment
Major components				
Stereo mix left				Track 1
Stereo mix right				Track 2
Perc. left flank	omni	left	3.5 m (11.5 ft)	Track 3
Perc. left main	cardioid	left	3.5 m (11.5 ft)	Track 4
Perc. right main	cardioid	right	3.5 m (11.5 ft)	Track 5
Perc. right flank	omni	right	3.5 m (11.5 ft)	Track 6
House left	cardioid	left	4 m (13 ft)	Track 7
House right	cardioid	right	4 m (13 ft)	Track 8
Microphones appearing only in the stereo mix				
String left flank	omni	left	3.5 m (11.5 ft)	
String left main	cardioid	left	3.5 m (11.5 ft)	
String right main	cardioid	right	3.5 m (11.5 ft)	
String right flank	cardioid	right	3.5 m (11.5 ft)	
Basses	cardioid	right	3 m (10 ft)	

an unknown quantity in terms of balance and relative levels, whereas the string ensemble was, by comparison, very stable and predictable.

Another important decision involving engineer, producer, and conductor was made regarding the relative perspectives of the string ensemble and the percussion ensemble. Physical constraints meant that the large array of percussion instruments would have to be placed *behind* the strings, but musical balance required that the two ensembles be essentially equal in all respects. In other words, they had to occupy the same apparent location on the recorded sound stage as the strings rather than be heard from their natural position behind the strings.

The solution was simple; it was to place an identical set of main microphones in front of each ensemble, treating them equally in terms of presence and level. Figure 16–11 shows the resulting sound stages for both stereo and surround presentation.

CONCERTO FOR PIANO AND STRINGS BY ALFRED SCHNITTKE

The recording was made on the large scoring stage at Lucasfilm's Skywalker Ranch in San Rafael, CA. Constantine Orbelian was both pianist and conductor of the Moscow Chamber Orchestra. The recording space has dimensions roughly of 27 m (90 ft) by 18.3 m (60 ft) by 9 m (30 ft), and its boundaries can be configured for a variety of acoustical requirements. For this recording the room was set for "moderately live" acoustics. Figure 16–12 shows details of studio setup and microphone placement. Note that the piano has been placed, with its cover

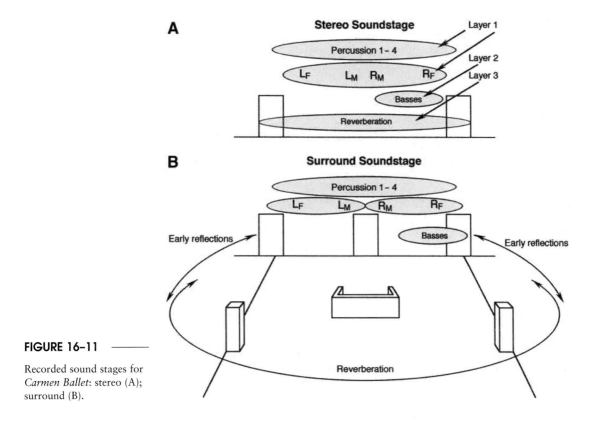

FIGURE 16-11

Recorded sound stages for
Carmen Ballet: stereo (A);
surround (B).

Studio Plan View

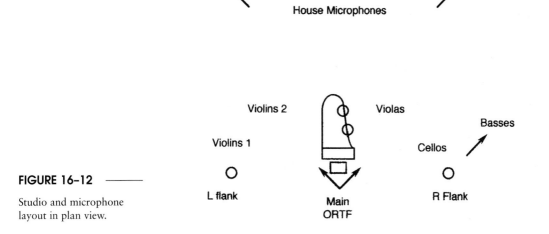

FIGURE 16-12

Studio and microphone
layout in plan view.

removed, so that it is well into the middle of the string group. Removing the cover was a necessity for ensuring eye contact among all players. Table 16–7 shows details of microphone and track deployment. Both stereo and surround sound staging are shown in Figure 16–13.

TABLE 16-7   Microphone and track deployment for Schnittke Piano Concerto

Position	Description	Stereo panning	Mic height	Track assignment
Major components				
Stereo mix left				Track 1
Stereo mix right				Track 2
Left ORTF	cardioid	left	3 m (10 ft)	Track 3
Right ORTF	cardioid	right	3 m (10 ft)	Track 4
Piano left	omni	left	2 m (80 in)	Track 5
Piano right	omni	right	2 m (80 in)	Track 6
House left	cardioid	left	3.5 m (11.5 ft)	Track 7
House right	cardioid	right	3.5 m (11.5 ft)	Track 8
These microphones appear only in the stereo mix				
Main left flank	omni	left	3 m (10 ft)	
Main right flank	omni	right	3 m (10 ft)	
Basses	cardioid	right	1.5 m (60 in)	

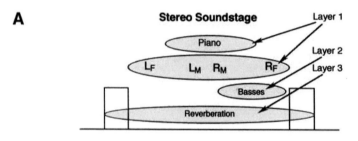

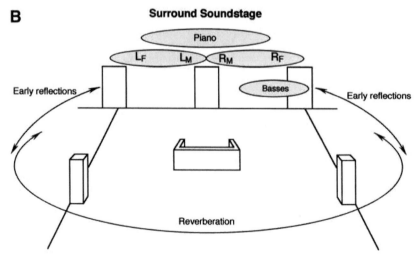

FIGURE 16-13 ————

Recorded sound stages for Schnittke Concerto: stereo (A); surround (B).

Because of fore–aft spacing limitations in the studio we introduced an additional 20 milliseconds of delay to the house microphones in order to "position" them as desired.

MUSICAL RECORDED REFERENCES

Hector Berlioz, "March to the Scaffold," *Symphonie Fantastique*, DVD Music Breakthrough, Delos International DV 7002, band 15.

Pyotr Tchaikowsky, *1812 Overture*, DVD Spectacular, Delos International DV 7001.

George Gershwin, *Rhapsody in Blue*, DVD Spectactular, Delos International DV 7002, band 12.

Hector Berlioz, *Te Deum*, DVD Spectacular, Delos International DV 7002, band 6.

Bizet-Schedrin, *Carmen Ballet*, DVD Music Breakthrough, Delos International DV 7002, bands 3–5.

Alfred Schnittke, *Piano Concerto*, Delos SACD 3259 (multichannel hybrid disc).

C H A P T E R 1 7

A SURVEY OF MICROPHONES IN BROADCAST AND COMMUNICATIONS

INTRODUCTION

The applications dealt with in this chapter include broadcast, news gathering, paging in public spaces, conference management, and safety alert systems. A number of microphone types used in these applications have already been discussed in Chapter 10, which covers microphone accessories. We refer to figures from that chapter where necessary as we underscore the specific applications in this chapter.

MICROPHONE TYPES USED IN BROADCAST AND COMMUNICATIONS

THE DESK STAND

The desk stand microphone mount is one of the oldest fixtures in communications. It long ago disappeared from broadcast in favor of the much more flexible pantograph assembly that allows the microphone to be supported on a heavy, stable base with the microphone itself positioned conveniently in front of the user. The desk stand, as shown in Figure 10–1, may still be found in a few boardrooms or paging systems, but in the vast majority of transportation terminals the telephone handset now takes the place of the desk stand.

THE TELEPHONE HANDSET

The telephone handset itself has undergone some important modifications. While the older carbon button transmitter (microphone) still has some advantages in normal telephony, it is often replaced by an electret element when the handset is to be interfaced with a paging or an announcing system. Care should be taken to use a handset/interface which will allow the system to mute before the handset has been placed back in its cradle, thus avoiding the disagreeable "thunk" of hanging up.

THE NOISE-CANCELING MICROPHONE

In very noisy environments, including airplane cockpits, ship decks, and heavy duty machinery rooms, a noise-canceling microphone is used, often in conjunction with a hands-free headset-microphone combination. The noise-canceling microphone is a gradient model designed for flat response at very short working distances. Stated differently, the normal proximity LF rise has been equalized for flat response relative to mid and HF, with the result that distant pickup is rolled off at LF. A typical model is shown in Figure 17–1A, and frequency response is shown at B. Close sound sources enter primarily by way of the front opening, while distant sound sources enter equally at front and back openings and are reduced 6 dB per halving of frequency below about 1 kHz. In the figure, the crosshatching shown at B indicates the effective range of noise cancellation. In order to gain this degree of effectiveness it is essential that the microphone be positioned virtually next to the talker's mouth. Such microphones are carefully designed with sufficient screening and mesh to attenuate close-in breath sounds.

BOUNDARY LAYER (BL) MICROPHONES

Positioning microphones very close to wall or floor boundaries has long been a general practice among knowledgeable engineers, but it was Crown International who, during the 1970s, introduced a line of microphones optimized for the purpose. These were known by the term PZM (pressure zone microphone), indicating that these microphones responded only to signal pressure components that were present at the boundary.

Boundary layer microphones are available from many manufacturers. While the early models were primarily omnidirectional, modern versions may have cardioid and hypercardioid patterns. When a directional pattern is used, its axis is parallel to the boundary layer and the microphone senses pressure gradient components that are parallel to the boundary. The BL microphone is often placed on a large room boundary or in the center of a large baffle, and when mounted in this manner the BL microphone picks up sound with a minimum of reflections. These microphones are ideal for placement on boardroom tables and altars, and have found a permanent home on-stage in legitimate theaters and concert halls. A typical omnidirectional model is shown in Figure 17–2A.

When placed on a supporting baffle, the larger the baffle the better the LF response, as shown in Figure 17–2B. When the baffle diameter is large with respect to wavelength, the microphone takes advantage of pressure doubling at the boundary surface. At progressively lower frequencies

A

B

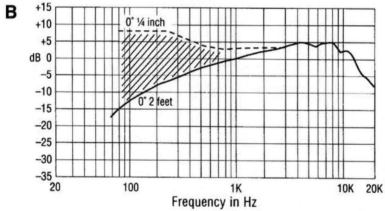

FIGURE 17-1

Photo of a noise-canceling
microphone (A); response
of the microphone to near
and far sound sources (B).
(Data courtesy of Crown
International.)

the boundary "unloads" and the microphone's sensitivity drops by 6 dB
as it approaches a free-space operating condition.

MICROPHONE GATING

For many applications an automatic microphone mixing and gating
system is used to limit the number of open microphones only to those
actually in use. (We discuss the electronic requirements for this in a later
section.) While conventional microphones are used in most of these
systems, Shure Incorporated has devised a mixing system in which the
microphones themselves can make a distinction in sound direction.
Figure 17–3 shows details of the Shure AMS26 microphone, which has

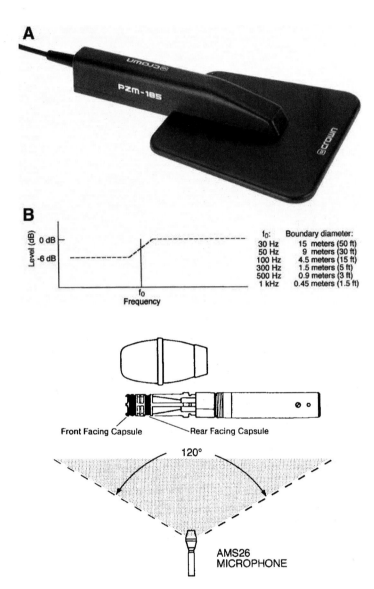

FIGURE 17-2

The boundary layer microphone (A); response depends on size of mounting baffle (B). (Photo courtesy of Crown International.)

f_0:	Boundary diameter:
30 Hz	15 meters (50 ft)
50 Hz	9 meters (30 ft)
100 Hz	4.5 meters (15 ft)
300 Hz	1.5 meters (5 ft)
500 Hz	0.9 meters (3 ft)
1 kHz	0.45 meters (1.5 ft)

FIGURE 17-3

Details of the Shure AMS26 microphone. (Data courtesy of Shure Inc.)

two elements, one forward-facing and the other rearward-facing. Only the front element is used in direct speech pickup; the other element is used only in a sensing circuit that allows the combination to reject random, distant sounds, while gating on frontal speech sound. It is obvious that the AMS type microphones must be used with AMS input circuitry for proper system performance.

HEADSET-MICROPHONE COMBINATIONS

On convention floors and in other active areas covered by radio and television broadcast, the roving reporter always wears a headset-microphone

FIGURE 17–4 ————

The headset-microphone combination. (Photo courtesy of AKG Acoustics.)

combination and carries a handheld vocal microphone for interviews. Three communications channels are in use here: two for the reporter and one for the interviewee. A typical unit with an electret microphone is shown in Figure 17–4.

MICROPHONES FOR TELECONFERENCING OPERATIONS

In teleconferencing, two or more remote locations are connected via telephone lines so that communication, both audio and low data-rate video, is enabled. Each of the remote spaces will need a center microphone that is broadly omnidirectional in the horizontal plane but limited in vertical coverage. Olson (1939) describes a microphone with a toroidal pattern, which is uniformly receptive to sound in the horizontal plane, but which has substantial rejection in the vertical direction. The microphone is of the interference and has polar response that varies with frequency. Cooper and Shiga (1972), as part of their research in quadraphonic sound, describe a relatively simple microphone with a toroidal pattern, as shown in Figure 17–5. The system consists of two figure-8 microphones oriented at 90° and placed one over the other. The two microphones are processed through all-pass networks with a 90° phase relationship between them, as shown.

HANDHELD MS MICROPHONE

The pistol-shaped microphone shown in Figure 17–6 is used for field newsgathering activities. It can be quickly re-aimed as needed. The M-section may be used alone for simple news gathering applications, while the

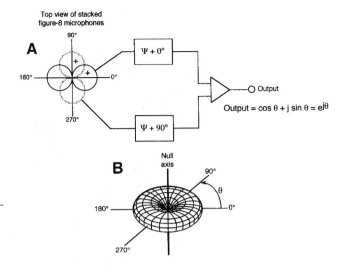

Top view of stacked
figure-8 microphones

Output = $\cos \theta + j \sin \theta = e^{j\theta}$

FIGURE 17-5

Conferencing microphone
with a toroidal directional
pattern.

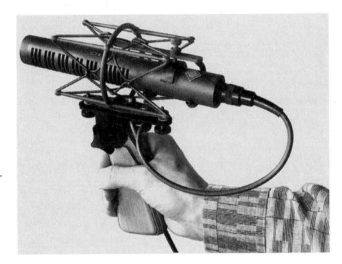

FIGURE 17-6

A handheld MS
microphone for news and
effects gathering. (Photo
courtesy of Neumann/
USA.)

MS combination is very useful for sound effects recording, where stereo remix in postproduction is often required.

THE TIE-TACK MICROPHONE

The present-day tie-tack microphone grew out of the relatively heavy lavalier microphone which hung around the user's neck. Those were also the days before wireless microphones, so the whole assembly was a clumsy one, with the user tethered by a cable. The heavy lavalier was also susceptible to noise pickup through contact with the wearer's clothing. The modern tie-tack microphone is very small electret model and can be clipped directly to the wearer's lapel or tie, as shown in Figure 17–7A. Details of the microphone itself are shown at B. The tie-tack microphone

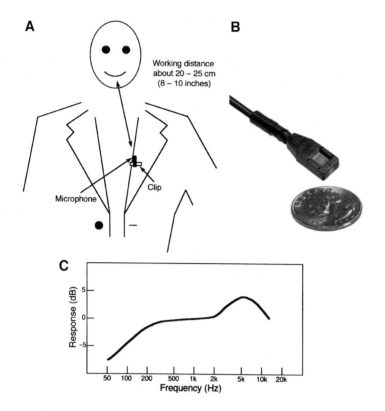

FIGURE 17-7

The modern tie-tack
microphone: typical
application (A); photo of a
typical model (B); normal
equalization of a tie-tack
microphone (C). (Photo
courtesy of Crown
International.)

is ubiquitous, barely visible in just about every talk or news show on
television. It is also the microphone of choice for lecturing in those situations where the talker must have freedom of movement. The response
pattern is normally omni, inasmuch as the operating distance is very
short. For feedback-prone environments a cardioid version can be used.
The microphone's response is normally internally equalized as shown at
C; the rolloff at MF compensates for the chest cavity resonances of the
talker, and the HF boost compensates for the microphone's position far
off-axis of the talker. Obviously, if a tie-tack microphone with a directional pattern is used, the main axis of the pattern must point toward the
talker's mouth.

HIGH-DIRECTIONALITY MICROPHONES

Sports events on TV make extensive use of telephoto shots of on-field
action, and it is desirable to have a microphone that can produce a matching sonic effect. Some of the larger rifle and parabolic microphones discussed in Chapter 6 are of some limited use on the playing field, as are
certain arrays discussed in Chapter 19. The "crack" of a baseball bat is a
great effect if it can be made loud and sharp enough. In many cases it has
been picked up with a small wireless microphone located at home plate.

IN-STUDIO ACOUSTICS AND PRACTICE

The announcer's booth is constructed as shown in Figure 17–8A. The space should be large enough for two persons so that interviews can be done comfortably. The acoustical treatment should consist of absorptive panels or other elements deep enough sufficiently to damp the lower frequency range of a male announcer (the range down to about 125 Hz), and the space should be free of any small room sonic coloration. Double

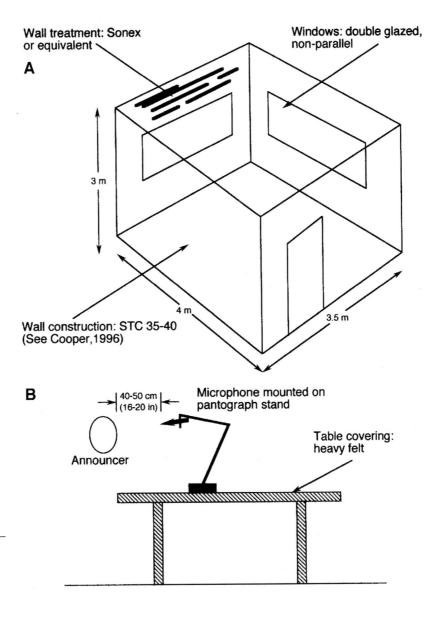

FIGURE 17–8

Radio announcer's booth: construction details (A); microphone usage (B).

glazing should be used in the windows of the booth that look into the control room and studio. The table in the center of the room should have a well-damped surface, such as felt, to minimize scuffing noises, shuffling of papers, and the like. We recommend that an acoustical consultant be engaged for advice regarding construction, room treatment, and the type of air ductwork necessary to minimize noise.

Details of microphone placement are shown at B. The engineer should be aware that modern broadcast practice relies a great deal on voice processing. Careful limiting, compression, and downward expansion are often used to maintain consistent levels and minimize noise. Additionally, some degree of equalization may be applied to the voice. Some announcers with "big, radio-type voices" may exhibit pronounced vocal waveform asymmetry, which may cause a high signal crest factor (see Chapter 2). The effect of this may be to restrict the maximum modulation level available for that announcer. Modern voice processors have circuitry that can be engaged to minimize high signal crest factors without audibly changing the timbre of the voice, thus providing for better overall on-air transmission.

CONFERENCE SYSTEMS

Large conferences more often than not are presented in midsize meeting rooms using a classroom setup. The basic requirement grows out of the fact that all attendees are participants and may be called upon to address the group as a whole. Often, several levels of priority are required. A chairperson, of course, is in charge of the meeting's proceedings. Additionally, sub-chairpersons may be responsible for smaller delegations. Here is a general description of such a system:

Chairpersons use a special station that allows them priority of all delegate stations of lower priority. Under normal conditions, when talking, the chairperson's voice is heard at all stations. When comments from the delegates are desired, the delegate engages his or her station; and this mutes the local station's loudspeaker so that feedback does not take place. Comments are then heard over the rest of the stations. The maximum number of units that can be engaged at a given time can be set by the chairperson, and the overall system gain is automatically adjusted downward in order to inhibit feedback. All the microphones are cardioid in pattern and are positioned, via the gooseneck, for fairly close placement to each delegate. In the system described here, up to 25 delegate stations can be connected in series-chain fashion, and a single base control unit can accommodate up to four such chains. An extension unit allows 75 additional delegate stations to be used. Figure 17–9A shows a photo of a typical user station that contains microphone, local loudspeaker, and control functions. A typical block diagram for system layout is shown in Figure 17–9B.

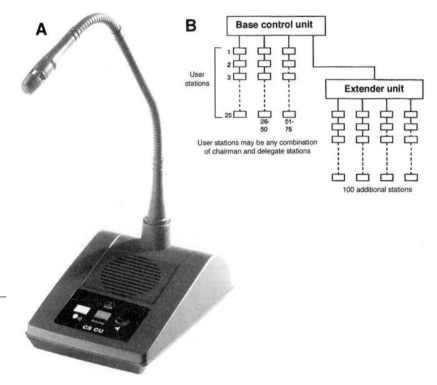

FIGURE 17-9

Conference systems: a typical user station (A); block diagram of typical system (B). (Data courtesy of AKG Acoustics.)

PAGING SYSTEMS

The modern airline terminal paging system is very complex. The telephone handset (with electret transmitter) is universally used for live announcements. Local stations exist at each gate, and the range of that system is limited, to the extent possible, to loudspeakers that specifically cover the local gate area. Most messages are obviously made by gate ticketing personnel and relate to the flight presently scheduled. Local gate systems operate best when they consist of a large number of small loudspeakers, each operated at a fairly low level and positioned uniformly throughout that gate area. This ensures minimum "spill" into adjacent gate areas.

Global paging is used for announcements that may affect a number of substations at one time, and messages are normally recorded in a "queue" system for replay a few seconds later. The necessity for the queue is to allow for a number of such announcements to "stack up," if need be, to be replayed in the order in which they were placed. This is a great convenience for the person making the page, since it ensures that the attendant can make the page immediately, get it into the queue, and then go on about their business.

At the highest level of priority are emergency pages or announcements, and these can override all local paging activities. Many such announcements are prerecorded and may be actuated by fire or other alarm systems. It is imperative that recorded announcements be made by professional announcers and appropriately processed (peak limited and spectrally shaped) for maximum intelligibility. In certain parts of the world it is also essential that such announcements be made in a number of languages.

The signal level of many paging systems is adjusted automatically and continuously over a range of several decibels, depending on the ambient noise level in various parts of the terminal as measured by sampling microphones placed throughout the terminal.

PAGING IN HIGH-NOISE AREAS

Paging in factories or on the deck of a ship is difficult. Noise levels are often so high that the excess levels over the speech range (250 Hz to 3 kHz) required to override the noise may be uncomfortable to all concerned. Intelligibility is actually reduced at such high levels. Today, key personnel on duty in noisy areas are often provided with pagers and cellular telephones for use in emergency situations; feeds from these channels may be routed directly, if needed, into audio communications channels.

CHAPTER 18

FUNDAMENTALS OF SPEECH AND MUSIC REINFORCEMENT

INTRODUCTION

It is rare to find any public venue today in which speech is not routinely reinforced. Today's auditoriums, worship spaces, classrooms, and arenas tend to be larger than in the past, and patrons have come to expect the comfort and ease of listening with minimum effort. In this chapter we will present an overview of sound reinforcement system design, with emphasis on those factors that determine the intelligibility of reinforced speech.

Music reinforcement spans the gamut from small-scale club systems to high-level reinforcement of rock and pop artists in major venues, both indoors and out. For some years, the emphasis in professional music reinforcement has been on basic sound quality, providing an impetus to both loudspeaker and microphone product development. We will begin our discussion with speech reinforcement.

PRINCIPLES OF SPEECH REINFORCEMENT

The basic requirements for a speech reinforcement system are that it provide:

1. Adequate and uniform speech signal levels for all listeners
2. Adequate speech intelligibility for all listeners
3. Natural speech quality for all listeners
4. Stable performance under all operating conditions

Before we discuss these requirements, let us look at the development of mass speech communication over the centuries:

In Figure 18–1A, a talker is addressing a group of listeners outdoors and must rely solely on the natural speech volume level to ensure that all listeners can hear. If the audience is a large one, those listeners in the

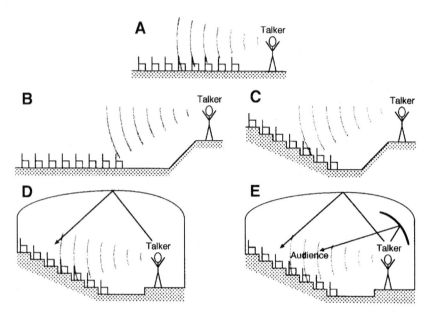

FIGURE 18-1 ————

The progress of assisted speech reinforcement over the centuries.

front will receive higher levels than necessary, while those at the back will have to strain to hear and understand the talker. Outdoor noises cannot be avoided.

In Figure 18–1B, the talker is elevated on a hillside. From this position the sound coverage of the audience will be more uniform, and more acoustical power from the talker will reach the audience.

In Figure 18–1C, the audience has been placed on a terraced area of an amphitheater, further isolating it from outdoor disturbances.

In Figure 18–1D, both audience and talker have moved indoors. The freedom from outside interference, along with the early reflections from the side walls and ceiling, will increase both sound levels and intelligibility.

The final step, shown in Figure 18–1E, is to improve the directivity of the talker with respect to the audience by adding a reflector behind the talker. This increases the level of the talker at mid-frequencies, further aiding intelligibility at a distance.

At each step, there has been an improvement in at least one of the following aspects: speech level, uniformity of coverage and minimizing outside disturbances. Until recent years, the development shown in Figure 18–1E was deemed sufficient to take care of virtually all speech requirements in small- to moderate-size lecture rooms.

ANALYSIS OF AN OUTDOOR (REFLECTION-FREE) SPEECH REINFORCEMENT SYSTEM

The analysis given here is based on the work of Boner and Boner (1969). Figure 18–2 shows the primary elements in a modern speech reinforcement system. There are four basic elements: talker, microphone,

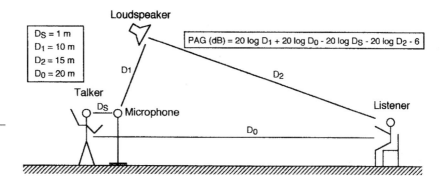

FIGURE 18-2 _____

Basic elements of an
outdoor speech
reinforcement system.

loudspeaker, and listener; these are separated by the distances D_S, D_0, D_1 and D_2, as shown.

Assume that the unaided talker produces an average speech level of 65 dB L_P at the microphone, 1 m (40 in) away. By inverse square law, the level at the listener will be 26 dB lower, or 39 dB L_P (see Chapter 2 under Acoustical Power). Both the loudspeaker and microphone are assumed to be omnidirectional.

When the system is turned on, the electrical gain of the microphone/loudspeaker combination can be increased until the loudspeaker provides a level from the microphone equal to that of the talker, or 65 dB. This condition produces unity gain through the system, and this results in acoustical feedback through the system as shown in Figure 18–3. Feedback causes the familiar "howling" effect that we have all heard with improperly controlled sound reinforcement systems.

To have a workable system it will be necessary to reduce the system gain by about 6 dB, which should result in an acceptable stability margin. When this adjustment has been made, the loudspeaker will now produce a signal level at 10 m (33 ft) (distance from the loudspeaker to the microphone) of 59 dB.

Now, we determine what level the loudspeaker will produce at the listener at a distance of 15 m (50 ft). Again, by inverse square law, we calculate that level to be 55.5 dB.

The acoustical gain of the system is defined as the difference in level at the listener with the system off compared to the level at the listener with the system on:

$$\text{Acoustical gain} = 55.5\,\text{dB} - 39\text{db} = 16.5\,\text{dB}.$$

A general equation for the *potential acoustic gain* (PAG) in dB that an outdoor system can produce is:

$$\text{PAG} = 20 \log D_1 + 20 \log D_0 - 20 \log D_s - 20 \log D_2 - 6 \qquad (18.1)$$

Later, we will see how the use of more directional microphones and loudspeakers can improve the maximum acoustical gain of a system.

Another graphical way of looking at the acoustical gain of the system is shown in Figure 18–4. We may think of the system as effectively

Physical circuit:

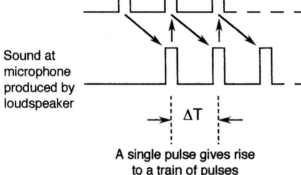

Microphone   Amplifier   Loudspeaker

Gain

Feedback path

Events in time:

Initial sound

Sound at
microphone
produced by
loudspeaker

ΔT

A single pulse gives rise
to a train of pulses

FIGURE 18–3

The origin of acoustical
feedback.

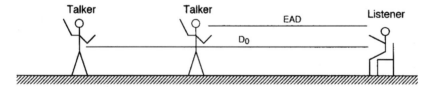

Talker   Talker   EAD   Listener

D_0

FIGURE 18–4

The concept of EAD.

moving the listener closer to the talker. When the system is off, the distance is of course D_0, or 20 m (66 ft). When the system is turned on, the talker has in a sense "moved forward" toward the listener. This "new listening distance" is known as the *equivalent acoustic distance* (EAD). We can calculate EAD by considering the following two facts: the talker produces a level of 65 dB at a distance of 1 m (40 in), and with the system turned on the listener hears the talker at a level of 55.5 dB. We now ask the question: at what distance from the talker will the level have dropped to 55.5 dB? The level difference is $65 - 55.5$, or 9.5 dB. Using the nomograph shown in Figure 2–7 we can see directly that level attenuation of 9.5 dB corresponds to a distance of about 3 m (10 ft), relative to a reference distance of 1 m (40 in). Therefore, with the system turned on, the listener will hear the talker as if that talker were located at a distance of 3 m (10 ft) from the listener.

ANALYSIS OF AN INDOOR SPEECH REINFORCEMENT SYSTEM

When we move the speech reinforcement system indoors its analysis becomes a bit more complicated. The inverse square attenuation with distance that we observed outdoors has now been replaced with a combination of inverse square loss and indoor reverberant level produced by room reflections.

An indoor system is shown in Figure 18–5A. We can see the inverse square paths to the listener, along with the reflected contributions from the room. When we look at the total contribution of both direct and reflected sound, the picture is as shown at B. Close to the source, the attenuation with distance follows the inverse square relationship; however, at some given distance, the reverberant field begins to dominate. The reverberant field is fairly uniform throughout the room, and at distances far from the loudspeaker the reverberant field is dominant. As we discussed in Chapter 2 (under The Reverberant Field) the distance from the loudspeaker at which both direct and reverberant fields are

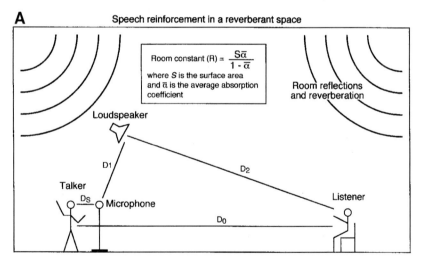

A Speech reinforcement in a reverberant space

Room constant (R) = $\dfrac{S\bar{\alpha}}{1 - \bar{\alpha}}$

where S is the surface area and $\bar{\alpha}$ is the average absorption coefficient

Room reflections and reverberation

Loudspeaker

D_1 D_2

Talker

D_S Microphone

Listener

D_0

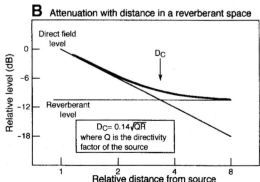

B Attenuation with distance in a reverberant space

Direct field level

D_C

Relative level (dB)

Reverberant level

$D_C = 0.14\sqrt{QR}$

where Q is the directivity factor of the source

Relative distance from source

FIGURE 18–5

An indoor speech reinforcement system: system layout (A); attenuation of sound with distance in a reverberant environment (B).

equal is known as critical distance (D_C). At D_C, the level is 3 dB higher than either of its components. In this example we are assuming that both talker and microphone are in the reverberant field produced by the loudspeaker.

A complete analysis of the PAG of an indoor system is fairly complex, but it basically takes the D terms in equation (18.1) and converts them to limiting values of critical distance. When this is done, two of the log terms in the equation cancel, and the net result is:

$$PAG = 20 \log D_{CT} - 20 \log D_s - 6 \text{ dB} \qquad (18.2)$$

where D_{CT} is the critical distance of unaided talker in the direction of the listener, and D_S is the talker-to-microphone distance.

The critical distance may be calculated by the equation:

$$D_C = 0.14\sqrt{QR} \text{ (meters or feet)} \qquad (18.3)$$

where Q is the directivity factor of the sound source and R is the room constant in the enclosed space. The room constant is:

$$R = \frac{S\overline{\alpha}}{1-\overline{\alpha}} \text{ (square units)} \qquad (18.4)$$

where S is the surface area in the space and $\overline{\alpha}$ is the average absorption coefficient in the space. (Note: the value of R is expressed in either square meters or square feet, depending on the system of units being used in the calculations.)

When we examine equation (18.2), we can see that the first term is dependent on the room characteristics and cannot easily be changed. Thus, the second term is the only one that we can easily change. This is intuitively clear; the best (and easiest) way to increase the potential gain of an indoor system is simply to move the microphone closer to the talker. In addition, the use of a directional microphone and employment of loudspeakers that aim the bulk of their output directly at the fairly absorptive audience will also improve system gain.

FACTORS THAT DETERMINE INDOOR SPEECH INTELLIGIBILITY

In an indoor environment there are three factors that largely determine how intelligible transmitted speech will be:

1. *Signal-to-noise ratio (dB)*. This is the ratio of average speech levels to the A-weighted local noise level. For best intelligibility, a signal-to-noise ratio of 25 dB or greater is recommended.

2. *Room reverberation time (s)*. When the reverberation time exceeds about 1.5 s, the overhang of successive syllables will have a detrimental effect on intelligibility. Strong room echoes in particular will have a deleterious effect on intelligibility.

3. *Direct-to-reverberant (D/R) speech level (dB).* When this ratio is less than about 10 dB, the reverberant level tends to mask speech in much the same way that random noise does.

In a later section we will discuss some of the methods by which the intelligibility of a speech reinforcement system may be estimated while system layout is still at the design stage.

MICROPHONES FOR SPEECH REINFORCEMENT

Figures 18–6 and 18–7 show some of the microphone types that are used in speech reinforcement. The hand-held vocal microphone is used as shown at Figure 18–6A. Vocal microphones are normally designed so that they produce fairly flat response when positioned about 5 to 10 cm (2 to 4 in) from the performer's mouth. (Typical response of a vocal microphone is shown in Figure 7–2.) At such small operating distances, the vocal microphone is fairly immune to feedback – a classic example of reducing D_S to a very low value. The best microphones for vocal use are those that have integral multiple screening surrounding the capsule to minimize the effects of inadvertent puffs of wind from the talker. Many of these microphones have a pronounced "presence peak" in the 3–5 kHz range for added brightness and improvement of articulation. Many performers feel very much at home with a vocal microphone in hand – and they often feel at a loss without it. Proper microphone etiquette must be learned; never blow on the microphone to see if it is on; always

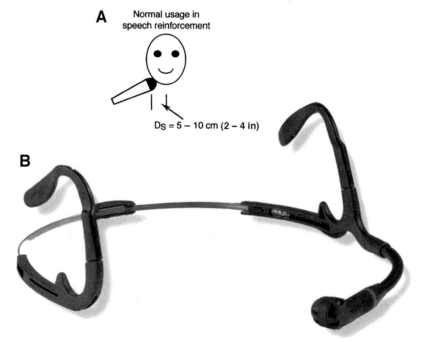

FIGURE 18–6

Microphones for speech reinforcement; the handheld vocal microphone (A); head-worn microphone (B). (Photo courtesy of AKG Acoustics.)

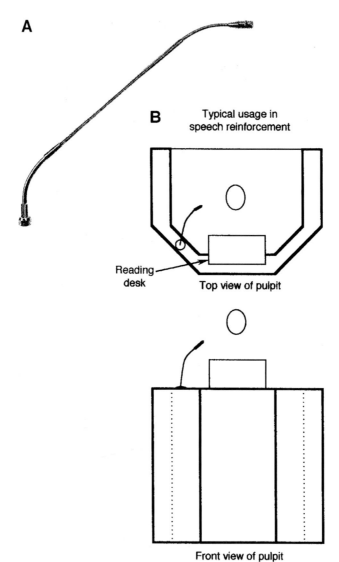

FIGURE 18-7 ————

Microphones for speech
reinforcement: podium
microphone (A); podium
usage (B); boundary layer
usage on an altar (C).
(Photo courtesy of Crown
International.)

hold it slightly to the side, outside the breath stream, and maintain
a consistent operating distance.

The head-worn microphone (shown at B) has long been a staple in
communications activities, but it was not until its adoption by singer
Garth Brooks that it became a staple for on-stage performance. The microphone element is an electret, normally with a cardioid or hypercardioid
pattern, and equalized for close use. When properly worn it is stable in its
positioning and offers excellent performance. It is invariably used with
wireless bodypacks, and for lecturers or panelists it provides complete
freedom of movement. Caution: it must be properly fitted and not be
allowed to slip into position in the breath stream of the talker or singer.

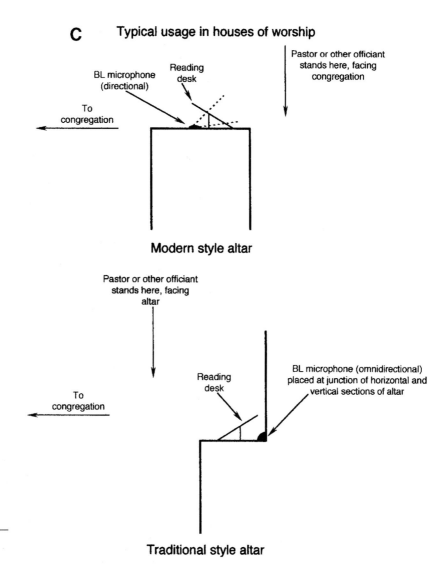

C   **Typical usage in houses of worship**

Pastor or other officiant stands here, facing congregation

BL microphone (directional)

Reading desk

To congregation

Modern style altar

Pastor or other officiant stands here, facing altar

To congregation

Reading desk

BL microphone (omnidirectional) placed at junction of horizontal and vertical sections of altar

FIGURE 18-7

Continued.

Traditional style altar

For permanent podium or lectern mounting there are numerous miniature electret cardioid or hypercardioid models mounted on flexible gooseneck extensions, as shown at Figure 18–7A and B. These can be unobtrusively located to one side and positioned at a distance of 30–50 cm (12–20 in) from the mouth of the talker. A small windscreen is recommended. It is important that the gooseneck portion be out of the range of movement of papers or notes used by the talker and that the talker's normal motions not be impeded.

For use on flat surfaces, such as tables used in panel discussions, or altars in houses of worship, a boundary layer microphone (shown at C) is essential. An omni pattern often works best, but cardioid models may be necessary to minimize local noises. The cardioid, however, will be more subject to impact noises than the omni. The operating distance is normally in the range of 45–60 cm (18–24 in).

The tie-tack microphone, which was introduced in the previous chapter, has the advantages of a small electret and is very popular primarily because it is inconspicuous. It is important that it be fastened to the user's lapel or tie with enough slack in the cable to avoid pulling or tugging on the cable as the performer moves around. For sedentary applications the microphone may be wired directly, but for most purposes it is used with a wireless bodypack. Position the microphone as high as possible on the tie or lapel; however, be aware that in a high position normal up and down head movements may cause audible shifts in level. Make the right compromise. The tie-tack microphone's response is normally shaped as shown to minimize radiation from the chest cavity and to maintain HF response.

TWO COMMON PROBLEMS

When microphones are positioned close to reflecting surfaces, the delayed reflection may combine with the direct sound, producing some degree of comb filtering in response. Figure 18–8A shows an omni

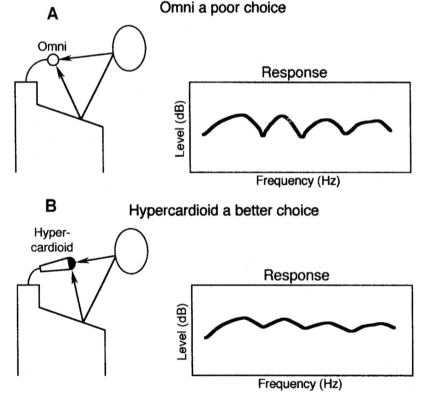

FIGURE 18-8

Delayed reflections; an omni microphone in the path of a reflection (A); a hypercardioid microphone with the reflection reduced by the off-axis attenuation of the pickup pattern (B).

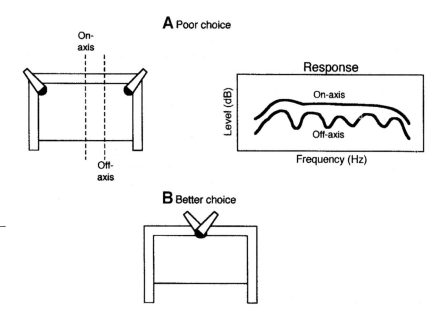

FIGURE 18-9

Cancellations due to multiple microphones; improper implementation (A); proper implementation (B).

microphone mounted on a podium in such a way that it will pick up a distinct reflection from the reading desk, producing uneven response. Moving the microphone to one side will alleviate this problem to some degree. A better solution is to use a hypercardioid microphone, whose off-axis response will attenuate the reflection, as shown at B.

Another common problem is the use of two microphones where one is sufficient. Improper usage is shown in Figure 18–9A. In broadcasting of important events, doubling of microphones is often done for transmission redundancy in case of failure of one channel, but more often than not both microphones end up operating in parallel. For a talker directly between the two there may be no problem. But talkers do move around, and the combined signals from both microphones will cause peaks and dips in response as shown. The solution is shown at B, where both microphones are mounted in coincident fashion and splayed slightly to increase the effective pickup angle. In this case the position of the talker will not be a problem since the talker's distance to the two microphones remains the same.

LOUDSPEAKER ARRAY OPTIONS IN LARGE SPACES

The basic loudspeaker approach for a venue depends on many things, but the reverberation time is paramount. Consider the space shown in Figure 18–10.

If the reverberation time is less than about 1.5 s, a single central loudspeaker array is usually the best solution. It has the advantage of single point radiation and will sound quite natural to all patrons. It must of course be designed so that it covers the audience seating area smoothly

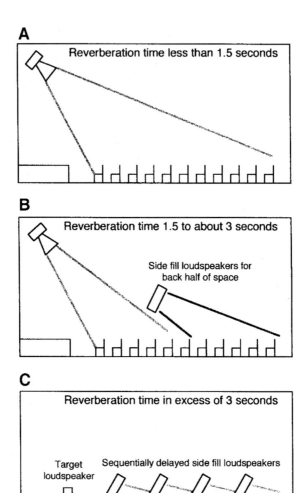

FIGURE 18-10 ⎯⎯⎯⎯

Speech reinforcement
in large spaces: low
reverberation time
(A); moderate
reverberation time (B);
high reverberation
time (C).

and minimizes sound projected onto the walls. The design approach is
shown at A.

If the same space has a reverberation time between about 1.5 and
3 s, then it may be more effective to supplement the central array with
additional side wall arrays at about halfway the length of the space. The
primary array is then adjusted so that it covers the front of the space, and
the secondary arrays are delayed so that the progressive wavefronts from
the main and secondary arrays will be effectively "in-step" at the back
of the space. The primary aim is to increase the D/R ratio in the back half
of the space. The design approach is shown at B.

Finally, if the same space has a reverberation time exceeding about
3 s, a conventional distributed system may be required. Here, there is no
central array as such, but rather a series of smaller loudspeakers located

on the side walls, sequentially delayed in order to produce a natural effect. In this approach, shown at C, each listener is relatively close to a loudspeaker and will thus benefit from the increase in D/R ratio. In many venues, these side mounted loudspeakers will be simple vertical column models, which have broad horizontal response (to cover the audience effectively) and narrow vertical response (to avoid excess signal "spill" onto the walls). Some installations include an undelayed *target* loudspeaker located at the front of the room which helps to define the actual sound source at that natural position.

As you can see, the design choices must be carefully made, and at each step in the design process a new analysis of the system's intelligibility estimate must be made. Very large venues such as sports arenas and stadiums are obviously more difficult to design and analyze.

Signal flow diagrams for the three systems of Figure 18–10 are shown in Figure 18–11. As the number of delay channels increases, the target coverage areas of loudspeakers in each delay zone are restricted to a specific portion of the seating area. The aim here is to provide coverage

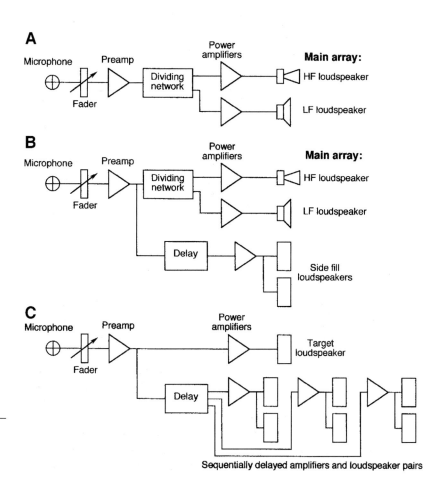

FIGURE 18-11 ———

Signal flow diagrams for the systems shown in Figure 18–10.

where it is needed for maximizing the D/R ratio. Ideally, the "spill" outside each area should be minimal, but in practice it is difficult to control.

ELECTRONIC CONTROL OF FEEDBACK

The usual cause of acoustical feedback, even in properly designed systems, is the careless raising of microphone input levels to "reach" for a weak talker. Another reason could be that too many microphones are open at the same time. When a skilled operator is at the controls, things rarely go wrong. But many systems are operated by amateurs or volunteers who do not understand the basic nature of feedback. In order to make systems failsafe in the hands of such operators, manufacturers have come up with a number of electronic devices that inhibit feedback to a greater or lesser degree. We discuss some of them below.

THE FREQUENCY SHIFTER

The frequency shifter was developed during the 1960s as a means of minimizing feedback by shifting the amplified sound up or down in frequency by about 4–6 Hz. As such, the effect was not easily noticed on speech. Music was however another matter; even slight frequency shifts are audible as a slow beating effect, especially in sustained passages. When the frequency is shifted, it is difficult for the conditions necessary for feedback to become established; however, excessive gain will result in a time-varying "chirping" effect as the system tries unsuccessfully to reach steady feedback. A signal flow diagram of an early frequency shifter is shown in Figure 18–12. Today, the technique is more sophisticated, consisting of a slow, random frequency shifting action rather than a fixed relationship.

NARROWBAND EQUALIZATION

In the hands of properly trained acousticians the insertion of narrowband notch filters into the audio chain can minimize feedback, resulting in improvements of up to 4 to 6 dB in overall gain capability (Boner and Boner, 1966). The technique involves driving the system slowly into feedback, determining the frequency of feedback, and inserting a narrowband filter at the feedback frequency. This procedure is done sequentially for the first three or four feedback modes of the system; beyond that

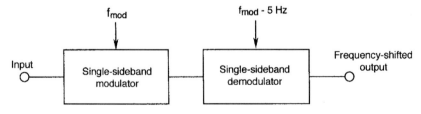

FIGURE 18–12

Details of a frequency shifter.

point there will be diminishing returns. The method is now normally carried out using parametric equalizer sections, as shown in Figure 18–13.

FEEDBACK ELIMINATORS

Based on digital signal processing, sophisticated methods of signal analysis and treatment can produce systems that can detect the presence of sustained feedback, determine the feedback frequency, and automatically engage the necessary filtering to control the feedback. A simplified signal flow diagram is shown in Figure 18–14.

AUTOMATIC MICROPHONE MIXING

Automatic mixers are widely used in many systems employing a number of microphones. For example, a house of worship may have microphones present on the lectern, pulpit, batistry, and altar. It is clear that only one of these microphones will be in use at a given time. Gating the microphones on and off does not necessarily require an operator; using a properly adjusted automatic microphone mixer, the control of the microphones can be smooth and foolproof. Figure 18–15 shows a basic signal flow diagram for an automatic mixer.

In many applications it will be necessary for more than one microphone to be open simultaneously, and under this condition there is a function in the mixer that will attenuate system gain according to the equation:

$$\text{Gain reduction} = 10 \log \text{NOM} \qquad (18.5)$$

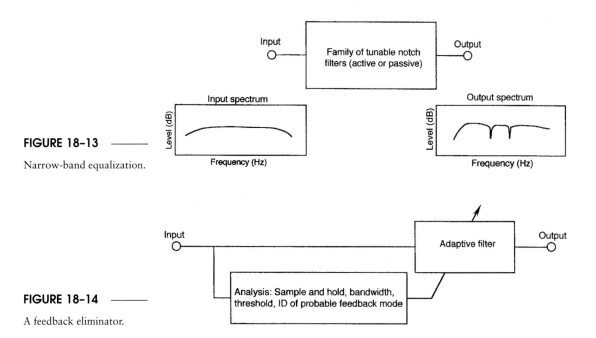

FIGURE 18-13 ———

Narrow-band equalization.

FIGURE 18-14 ———

A feedback eliminator.

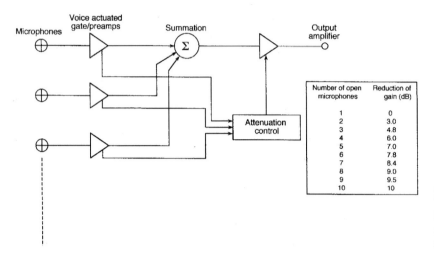

FIGURE 18–15

Basic signal flow
diagram for an automatic
microphone mixer.

where NOM is the *number of open microphones*. The desired gain
reduction amounts to −3 dB for each doubling of open microphones and
will result in a uniform reinforced level in the space.

A well-designed automatic mixer will also provide some immunity
against false triggering of gating functions due to variations in the ambient noise level of the space.

ESTIMATING THE INTELLIGIBILITY OF A SPEECH REINFORCEMENT SYSTEM

Where a speech reinforcement system has been installed in a venue, the
effective intelligibility of that system can be determined directly by syllabic testing using standardized techniques. Essentially, a talker stands at
the microphone and reads a set of random syllables, all embedded in a
"carrier" sentence. An example of this is: "Please write down the word
cat;" now I want you to write down the word *man*." And so it goes. The
purpose of the carrier sentence is to present the test syllables within the
acoustical masking context of continuous speech. The results for various
listeners in various parts of the auditorium are then analyzed and the
accuracy of their responses expressed as a percentage. If a listener gets
a score of 85% on random syllabic testing, then that listener will likely
understand about 97% of normal speech in that space.

If a speech reinforcement system is still on the design drawing board,
its effectiveness may be broadly estimated, based on certain acoustical
parameters that the supervising acoustician can arrive at. For example,
the acoustician can estimate the normal direct speech sound levels at
a typical listener that will be produced by the targeted loudspeakers. The
acoustician can also determine, based on reverberation analysis, what
the reverberant level at the listener will be, as well as the reverberation
time itself. The acoustician can also make a reasonable estimate of the

noise level likely to be encountered in the space, based on the intended degree of noise isolation and elimination.

If all of these estimates can be made for the octave frequency band centered at 2 kHz, and if it can be safely assumed that there are no unusual reflections in the room's reverberant decay pattern, then an estimate of the system's articulation loss of consonants ($\%Al_{cons}$) can be made using the following set of equations:

$$\%Al_{cons} = 100 \times (10^{-2(A + BC - ABC)} + 0.015) \qquad (18.5a)$$

$$A = -0.32 \log\left(\frac{E_R + E_N}{10\ E_D + E_R + E_N}\right) \qquad (18.5b)$$

$$B = -0.32 \log\left(\frac{E_N}{10\ E_R + E_N}\right) \qquad (18.5c)$$

$$C = -0.5 \log\left(\frac{T_{60}}{12}\right) \qquad (18.5d)$$

where

$$E_R = 10^{L_R/10}$$
$$E_D = 10^{L_D/10}$$
$$E_N = 10^{L_N/10}$$

As an example, assume that our preliminary room and system simulations give the following values: $T_{60} = 4$ seconds, $L_R = 70$ dB, $L_D = 65$ dB and $L_N = 25$ dB. Calculating the values of A, B and C:

$$A = 0.036$$
$$B = 1.76$$
$$C = 0.24$$

Entering these values into equation (18.5a), gives $\%Al_{cons} = 14\%$.

Figure 18–16 indicates a subjective assessment of this system's anticipated performance as on the borderline between adequate and poor.

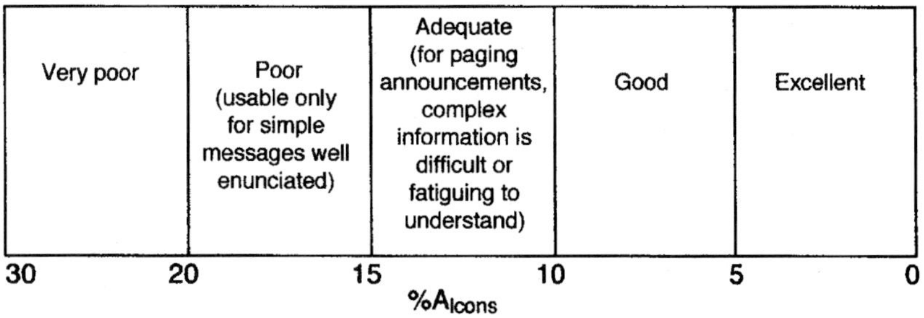

Very poor	Poor (usable only for simple messages well enunciated)	Adequate (for paging announcements, complex information is difficult or fatiguing to understand)	Good	Excellent
30	20	15	10	5	0

FIGURE 18-16 ———

Subjective descriptions of $\%Al_{cons}$ values.

%Al_cons

Based on this observation, the acoustician and architect may find common ground between them for improving the estimate. Most notably, a reduction of reverberation time (difficult, and often costly, to accomplish) or an improvement in loudspeaker coverage for higher direct sound levels (easier to accomplish) might be considered.

SPEECH AND MUSIC REINFORCEMENT IN THE LEGITIMATE THEATER

Traditionally, legitimate theaters and playhouses have been fairly small spaces, and professional actors have long been known for their ability to fill a house without electroacoustical assistance. This is still the model in most established venues. However, modern venues tend to be larger than earlier ones, and the proliferation of multi-purpose halls can result in staging of drama in houses that are actually too large for the purpose. In these cases a speech reinforcement strategy as shown in Figure 18–17 is very effective. Here, three boundary layer microphones are placed on the apron of the stage as shown. Ideally, these microphones should be in the direct field of the actors, but in some large houses the microphones may be in the transition region between direct and reverberant fields.

The signals are routed to carefully equalized and delayed loudspeaker channels located in the proscenium. In many cases, the acoustical gain of the system at mid-frequencies may be virtually none; the important effect of the system is the amplification of high frequencies, which enhances intelligibility. The purpose of the delay is to ensure that the first-arrival sound at the listeners is from the stage.

If omni microphones can provide the requisite gain at high frequencies, they are the first choice. If cardioid microphones are needed for

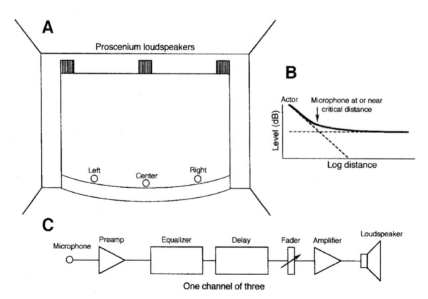

FIGURE 18–17

Stereophonic reinforcement of speech in the legitimate theater: stage layout (A); actor-microphone relationship (B); typical electrical channel (C).

feedback minimization, be aware that footfalls on the stage may become a problem, which will require high-pass filtering.

Special effects and off-stage events are normally handled by a sound designer and may be fed into the house from a variety of directions.

THE MODERN MUSICAL THEATER

Since the rock musicals of the 1960s, sound reinforcement has become a requirement in the musical theater, both for pit musicians and singer–actors on stage. Today, hardly a touring company goes on the road without a sound specialist and all the necessary equipment to do the job right. Here are some of the microphone requirements:

1. *On-stage*: Singers wear wireless tie-tack microphones, usually positioned at the middle of their hairlines. This affords complete freedom of movement and expression. See the notes in Chapter 9 regarding the use of large numbers of wireless microphones.

2. *Overhead pickup*: Rifle microphones may be flown to pick up some degree of stage ambience or group action.

3. *Orchestra pit*: The pit orchestra has always been limited in personnel, and individual clip-on microphones may be used on each instrument. These signal outputs must be pre-mixed to give the front-of-house (FOH) mixer the necessary flexibility in handling the overall production. Assistant mixers may be needed in complex situations.

Loudspeakers are normally deployed in vertical columns at the sides of the stage, with other loudspeakers positioned in the house as necessary for special effects.

HIGH-LEVEL CONCERT SOUND REINFORCEMENT

The mega-event concert performance of a major pop/rock artist is usually presented in a large indoor arena, outdoor stadium, or even an expansive field. Patrons pay dearly for front seats, and even those who are seated at great distances will expect sound pressure levels in the 110–115 dB range.

Most of the instruments have direct feeds to the console, and only vocals, drums and an occasional wind instrument will actually require microphones. All microphones are positioned very close to the sound sources to minimize feedback since on-stage monitoring levels can be very high. Vocalists–instrumentalists in particular have gravitated to head-worn microphones, since these give freedom of movement as well as a very short and consistent operating distance for vocal pickup. A solo vocalist may still prefer a handheld microphone since this has long been a standard "prop" onstage.

A potential source of feedback is from stage floor monitor loudspeakers operating at high levels into the performer's microphones, even when those microphones are operated at very short distances. As a hedge to this, a new technique known as *in-the-ear monitoring* has become popular. It uses small receivers that are actually placed in the wearer's ear canal, with operating levels carefully monitored.

ACTIVE ACOUSTICS AND ELECTRONIC HALLS

The field of active acoustics has been around for at least three decades and has made steady but gradual progress. One of the earliest systems was *assisted resonance* (Parkin, 1975). The first notable installation was in London's Royal Festival Hall. The hall lacked sufficient volume for its seating capacity and target reverberation time, and the system was installed to increase reverberation time in the range 60–700 Hz. The system consists of 172 channels. Each channel contains a microphone placed in a Helmholtz resonator, amplifier, and a loudspeaker. The microphones are positioned well in the reverberant field. Because of their narrow frequency bands the individual channels are quite stable, and the increase in apparent reverberation time results from the high-Q nature of each resonator. Figure 18–18A shows a view of a single channel, while the increase in reverberation time, with the system on and off, is shown at B.

LARES stands for Lexicon Acoustical Reverberance Enhancement System. The system consists of a set of Lexicon reverberation channels randomly feeding an ensemble of loudspeakers. Microphones are located above and in front of the performance area. The increase in reverberation time is the result of the reverberation generators themselves, and the gain of the systems is stabilized by randomly varying delay sections in

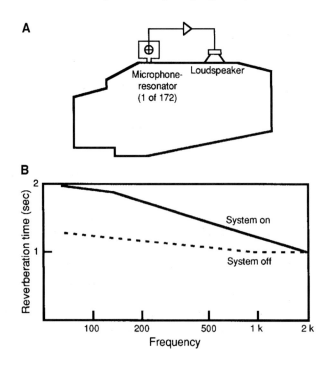

FIGURE 18–18

Parkin's assisted resonance.

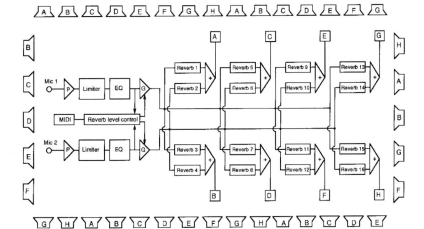

FIGURE 18-19

The LARES system.

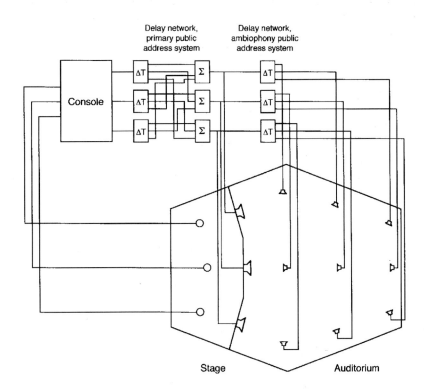

FIGURE 18-20

Delta stereophony.

each reverberation channel. Reverberation time is independently adjustable and is not dependent on reverberant level. A signal flow diagram for the system is shown in Figure 18–19.

Delta stereophony, shown in Figure 18–20, is used to increase loudness of stage events without compromising the natural directional cues from the stage. The system includes loudspeakers positioned at stage level as well as overhead. The number of delay channels normally does not exceed about six or eight, and the system is easily reconfigured for various kinds of stage events.

C H A P T E R   1 9

OVERVIEW OF MICROPHONE ARRAYS AND ADAPTIVE SYSTEMS

INTRODUCTION

A microphone array may be considered to be a set of microphone elements arrayed in space whose outputs are individually processed and summed to produce a given output. In a basic sense, an early cardioid microphone created by summing separate pressure and gradient elements may be thought of as an array, but this is not what we have in mind. The arrays and techniques discussed here normally use multiple elements to achieve a target directional response that may be specific for a given application. In other cases, signal processing may be applied to single microphones, or microphone channels, to enhance performance in some unique respect.

Adaptive systems are those whose signal processing coefficients change over time in order to accommodate or maintain a given signal control parameter. Basic adaptive techniques will be discussed as they apply to echo cancellation and as they are used to simulate, via data reduction techniques, the output of microphones in recording applications. Other applications of adaptive techniques deal with the manipulation of array directivity as a function of source tracking or other requirement. The discussion begins with basic line array theory.

DISCRETE LINE ARRAY THEORY

The simplest line array consists of a group of equally spaced microphones whose outputs are summed directly. Let us assume a set of four omnidirectional microphones arrayed vertically, as shown in Figure 19–1. Let the spacing, d, between adjacent line elements be 0.1 m

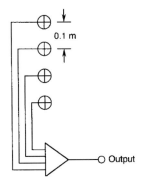

0.1 m

Output

FIGURE 19–1 ————

A four-element in-line array of omnidirectional microphones.

(4 in). The far-field directivity function $R(\phi)$ is given by the following equation:

$$R(\phi) = \frac{\sin (1/2\ Nkd \sin \phi)}{N \sin (1/2\ kd \sin \phi)} \tag{19.1}$$

where $N = 4$ and $d = 0.1$. The measurement angle ϕ is given in radians, and $k = 2\pi f/c$, where c is the speed of sound in m/s.

Figure 19–2 shows directivity plots in the vertical plane for the four-element array at frequencies of 400 Hz, 700 Hz, 1 kHz and 2 kHz. A plot of the directivity factor of the array is shown in Figure 19–3 by

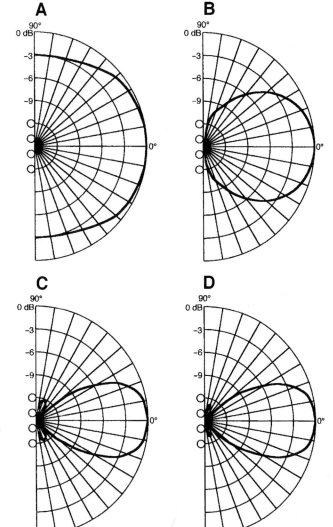

FIGURE 19–2 ————

Far-field directional response for a four-element array of omnidirectional microphones with inter-element spacing of 0.1 m: 400 Hz (A); 700 Hz (B); 1 kHz (C); 2 kHz (D).

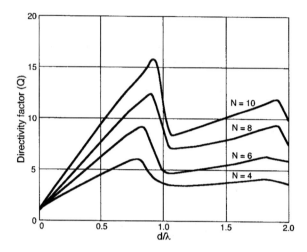

FIGURE 19-3

Directivity factor for in-line arrays of 4, 6, 8 and 10 elements.

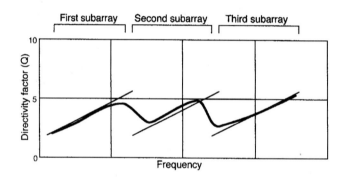

FIGURE 19-4

Extending the frequency range of uniform coverage.

the curve marked $N = 4$. The value of $d/\lambda = 1$ along the baseline of the graph corresponds to a frequency of 3340 Hz.

The useful range of the array extends up to a value of d/λ equal to about 1. Above that point the directional pattern shows the development of numerous off-axis lobes, although the on-axis directivity factor remains fairly uniform. The useful range of the array can be extended by shortening it with rising frequency. This involves having numerous elements and crossing over from one wider-spaced set to another with smaller spacing with increasing frequency. With careful attention to matters of frequency division, response as shown in Figure 19–4 is possible.

A CONSTANT-DIRECTIVITY TRANSDUCER ARRAY USING LOGARITHMICALLY SPACED ELEMENTS

The frequency range of a directional array using logarithmically spaced elements is more efficient in the use of a fixed number of transducers than an array consisting of equally spaced elements. Van der Wal et al.

FIGURE 19-5 ————

A logarithmically spaced line array of microphones (Van der Wal et al., 1996).

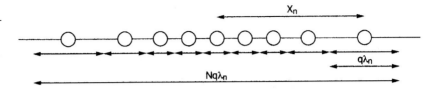

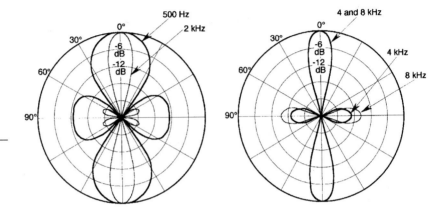

FIGURE 19-6 ————

Polar response estimates for the array shown in Figure 19–5 (Van der Wal et al., 1996).

(1996) propose an array as shown in Figure 19–5. The spacing of elements is more dense toward the center of the array, and this allows a decreasing array size by filtering with rising frequency such that the number of active transducers is optimized in all frequency ranges. The directivity of the array is shown in Figure 19–6. Here, the measured directivity over the four-octave range from 500 Hz to 8 kHz shows a remarkably uniform central lobe, with minor lobes largely in the -20 dB range.

MICROPHONE ARRAY FOR MULTIMEDIA WORKSTATIONS

Mahieux et al. (1996) describe an array consisting of 11 elements placed in a slight arc above the display screen of a multimedia workstation. The operating distance from the microphones to the operator is about 0.7 m, well in the near field. The purpose of the array is to pick up the voice of the operator with a fairly uniform directivity index of 10 dB over the frequency range from about 500 Hz to 8 kHz. The elements of the array are grouped into four subarrays as shown in Figure 19–7. Through appropriate filtering into four frequency bands a net directivity index is produced as shown in Figure 19–8.

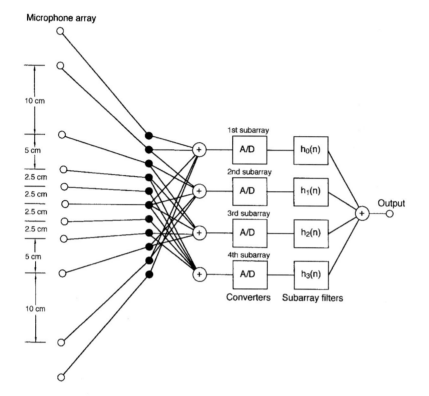

FIGURE 19–7

Top view of line array built into a multimedia workstation (Mahieux et al., 1996).

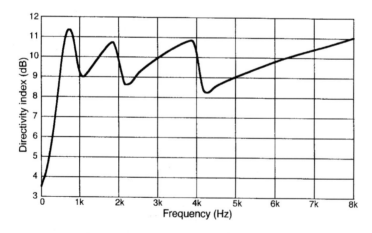

FIGURE 19–8

Directivity index of the array shown in Figure 19–7 (Mahieux et al., 1996).

A DESK-TYPE MICROPHONE ARRAY WITH CARDIOID HORIZONTAL RESPONSE AND NARROWED VERTICAL RESPONSE

The German Microtech Gefell model KEM970 microphone is an example of a fixed vertical array that produces a broad lateral cardioid pickup pattern and a 30° (−6 dB down) vertical pattern over the frequency range

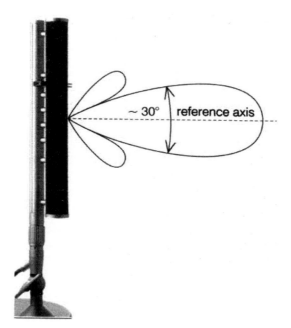

FIGURE 19-9 ———

Microtech Gefell KEM970
microphone. (Photo
courtesy of Cable Tek
Electronics Ltd.)

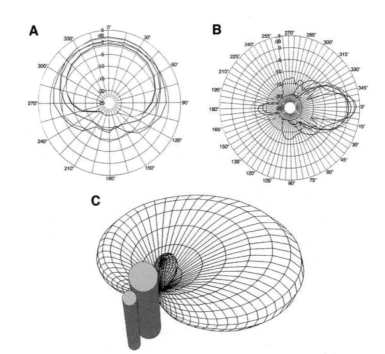

FIGURE 19-10 ———

Microtech Gefell KEM970
microphone polar data:
horizontal polar family (A);
vertical polar family (B);
typical directivity balloon
(C). (Data courtesy of
Cable Tek Electronics Ltd.)

above 800 Hz, resulting in a directivity index of about 10 dB at MF and
HF. The length of the array is 0.35 m (14 in). Figure 19–9 shows a side
view of the microphone, and Figures 19–10A and B show families of
polar response. A typical "directivity balloon" for the microphone is
shown at C.

AN ADAPTIVE HIGH-DIRECTIONALITY MICROPHONE

The Audio-Technica model AT895 microphone is shown in Figure 19–11, and a view of its elements is shown in Figure 19–12. The center element is a short-section line microphone of the type discussed in Chapter 6, and the four capsules at the base are cardioids. In normal operation, opposite pairs of cardioids are subtracted to produce a pair of figure-8 patterns at MF and LF which are oriented 90° to each other. These resulting patterns are perpendicular to the line element and thus do not pick up sound along the main axis of the microphone. They do of course pick up sound at MF and LF arriving at right angles to the line element. The signal flow diagram of the system is shown in Figure 19–13.

The adaptive portion of the system, which is digital, derives signal correlation values among the three outputs from the microphone elements and determines, on a continuing basis, the amount of interfering sound level at MF and LF. The interfering signals are added to the output of the line element in reverse polarity so that cancellation takes place at MF and LF. Much of the art in development of the system was in determining the range of operation, both in level manipulation and frequency, and also the time constants necessary for smooth, unobtrusive operation.

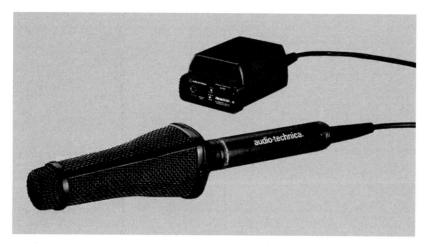

FIGURE 19–11 ———

Photo of Audio-Technica model AT895 adaptive microphone. (Photo courtesy of Audio-Technica US, Inc.)

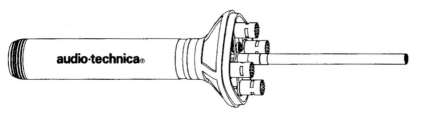

FIGURE 19–12 ———

Positions of microphone elements. (Figure courtesy of Audio-Technica US, Inc.)

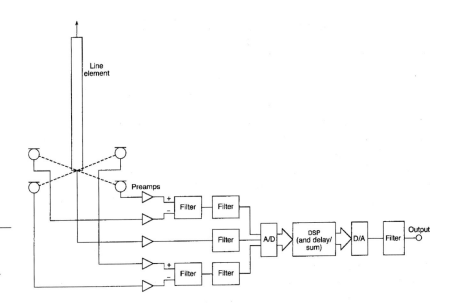

FIGURE 19–13

Signal flow diagram
for AT895 adaptive
microphone. (Figure after
Audio-Technica US, Inc.)

There are three basic modes of system operation:

1. *All elements operating.* This produces maximum rejection of interfering signals.
2. *Line element plus one figure-8 set.* This produces rejection only in one pickup plane; useful under conditions where interfering sources are largely confined to the ground plane.
3. *Line element operating alone, with optimum filtering.*

The microphone is intended primarily for activities such as news gathering and sports events, where interfering sounds may be constantly changing in directional bearing.

ECHO CANCELLATION

Modern conference systems allow users in one location to communicate easily with another group at some distance. *Duplex operation* of the systems permits conversation to proceed freely in both directions and is normally a basic requirements of these systems. A fundamental problem exists in a simple system, such as that shown in Figure 19–14. If talkers in one location are communicating with a distant location, they will hear their own voices delayed by the cumulative characteristics of the transmission path as the signal passes through the loudspeaker-microphone pair at the receiving end and is returned to the sending location.

These "echoes" may occur in a fraction of a second – or they may, as in the case of satellite links, be in the range of a second or longer. In addition, the return signal will exhibit noise as well as the acoustic

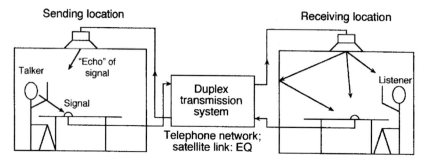

FIGURE 19-14

A basic duplex (two-way) conferencing system: signals originating at the sending location are reproduced at the receiving location via a loudspeaker; the return signal heard by the original talker will be delayed by the transmission path and further contaminated by room reflections at the receiving end.

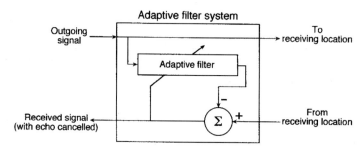

FIGURE 19-15

Adaptive filtering can be used to cancel the "echo" of the original signal from the sending end; the adaptive filter system intercepts the original signal and compares it with the return signal; any differences are forced to zero via a negative feedback path, and the signal returned to the sender is thus free of the echo.

signature of the receiving location. There may be equalization changes due to a variety of other transmission effects and, taken as a group, these effects will be quite disturbing to anyone attempting to communicate over the system.

Adaptive filtering can be used to alleviate these problems, and operation of the filter is shown in Figure 19–15. Because it is a generic filter, it can provide delay functions along with equalization. The filter is located at the sending end of the system and actually "models" the downstream transmission and return paths, comparing their output with the original signal from the sending location. The filter is placed in a negative feedback loop that actively "seeks" to reduce the difference between the original and received signals to zero, thus cancelling the echo.

In modern systems, the initialization of the system is an automatic process and usually takes place in a matter of a second or so after the

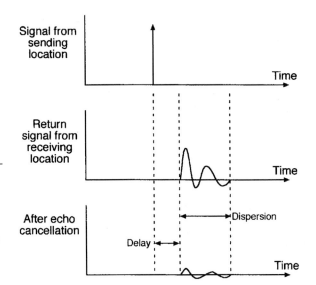

FIGURE 19-16

The original signal is returned with delay (echo) and further acoustical contamination; the adaptive filter cancels these effects to a large degree. (Data presentation after Aculab.)

system is turned on. Once initialized, the system remains fairly stationary; updating itself only when there is a substantial change in the overall signal path, such as moving the microphone, opening or closing doors, and the like, at the receiving end.

Figure 19–16 shows typical action of the adaptive filter. The originating signal is modeled as an impulse at A, and the cumulative effects of round-trip transmission are shown at B. The cancelled signal is shown at C. Note that the direct-through path from the microphone at the receiving end is outside the filter's feedback loop and is fed directly back to the loudspeaker at the sending end. Thus, the communication path remains open at both ends at all times.

Modern systems are quite stable, provide more than adequate gain, and can handle a number of microphones through automatic gating. Woolley (2000) provides an excellent overview of the subject.

"VIRTUAL" MICROPHONES

Kyriakakis et al. (2000, 2002) describe the concept of a *virtual microphone* in the context of a classical orchestral recording in which the outputs of selected microphones may be encoded at a low bit rate and reconstructed later as needed. Figure 19–17 shows a partial layout for an orchestral recording session; there are two main microphones (an ORTF pair) and a single accent microphone for timpani pickup. (In actuality there would be many more microphones, but for our discussion here we will consider only the main pair and a single accent microphone.)

We will record the primary ORTF microphone pair at the full data rate, but the timpani microphone is to be recorded as a virtual microphone. The timpani microphone signal will be a subset of the left ORTF

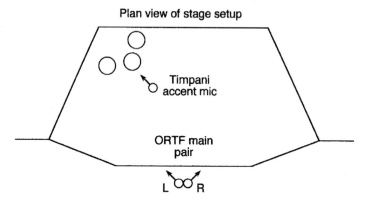

Plan view of stage setup

Timpani
accent mic

ORTF main
pair

L ∞ R

FIGURE 19-17 —————

Basic layout for a classical
recording session; only the
main microphone pair and
a single accent microphone
are shown for clarity.

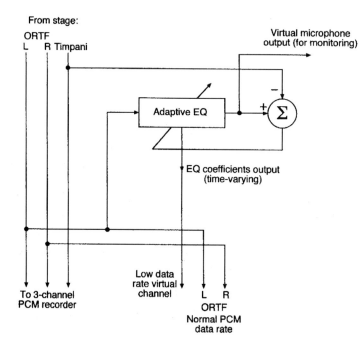

From stage:
ORTF
L   R Timpani

Virtual microphone
output (for monitoring)

Adaptive EQ

−
+ Σ

EQ coefficients output
(time-varying)

To 3-channel
PCM recorder

Low data
rate virtual
channel

L   R
ORTF
Normal PCM
data rate

FIGURE 19-18 —————

A normal digital (PCM)
recording path shown at
the left in the figure; the
path shown at the right
stores the virtual timpani
track as a time-varying set
of adaptive equalization
coefficients, which can later
be convolved with the left
ORTF signal to produce a
reconstruction of the
original timpani signal.

channel signal. This subset consists of filter coefficients that define, on a moment by moment basis, the output of the timpani microphone in terms of the overall spectrum picked up by the left ORTF microphone. The left ORTF channel has been chosen for this purpose since its pattern and orientation contain substantial signal from the timpani.

On playback, the virtual output of the timpani microphone will be recovered by time-varying equalization of the left ORTF signal via an adaptive filter. The filtering generates low data rate coefficients which are stored during the recording operation, resulting in a continuous "re-equalization" of the left ORTF signal so that its spectrum is identical to that of the actual timpani microphone signal. This process is shown in Figure 19–18.

The effectiveness of the entire process depends largely on instantaneous musical balances as they are picked up by the ensemble of all microphones. If thresholds of the virtual signal are high enough in the main microphone channels, then the recovery can be excellent. The signal recovered in this manner can then be used for postproduction rebalancing of the program.

Even in this age of high data rate delivery media, the requirements of extended multichannel systems will at some point strain the available hardware resources. The virtual microphone principle offers many attractive solutions: for example, large venue surround sound presentations can be outfitted with a large number of ambience channels, each one corresponding to a microphone positioned in the original recording venue at various distances, near and far, from the orchestra. If these signals are recorded via virtual microphones the overall data rate can be vastly reduced.

FINAL NOTES

In this chapter we have covered only a few of the many areas where microphone arrays and adaptive signal processing are useful. The interested reader will want to explore the following application areas:

1. *Hearing aids*: Taking advantage of binaural hearing, basic beam-forming techniques can lock in on nearby sources, providing the hearing-impaired user with stronger localization cues and better speech intelligibility. Adaptive beam-forming can be used to alter the directivity patterns at each ear, resulting in similar improvements.

2. *High-directivity beamforming*: As with loudspeakers, large arrays of microphones can be designed for high directivity over a fairly wide frequency range. If the arrays are adaptive they can be used for tracking individual sources over large angular ranges in difficult environments.

3. *Blind deconvolution*: In situations where multiple signals of unknown source location are presented simultaneously, advanced techniques can be used, within broad limits, to sort out the signals.

An excellent starting point for further studies is the book *Microphone Arrays* (Brandstein and Ward, 2001).

CHAPTER 20

CARE AND MAINTENANCE OF MICROPHONES

INTRODUCTION

There is relatively little that sound engineers and technicians can do in the way of actual maintenance of microphones, outside of minor external repairs. Far more important are the routine things that can be done on a daily basis to care for them and to ensure that they are working properly. All technical personnel in a given broadcast, recording, or sound reinforcement activity should be taught the rudiments of proper microphone handling, and if these rules are followed a microphone should last virtually indefinitely with little or no departure from its original response. In addition, there are number of easy measurements and performance comparisons that may determine if a microphone is performing according to its specifications.

GENERAL RECOMMENDATIONS

When not in actual studio use, microphones should be stored in the boxes in which they were shipped. A good alternative here is to construct a cabinet with felt-lined sections for each microphone, and it is recommended that the microphone clip, or mount, normally used with the model be stored in the same slot. Fussy managers will demand that the cabinet be locked at all times and that operating personnel be given keys. Many older models of dynamic and ribbon microphones have open magnetic structures with a considerable stray magnetic field that can attract ferric dust particles. Over the years this dust can pile up and possibly impair performance, so be on the lookout for it.

Handheld microphones take the greatest abuse in any location. Inexperienced users may inadvertently drop them, and prolonged close-in use will result in breath moisture, even saliva, on the protective screen. It is a good idea to wipe off all handheld microphones after use with a moistened lint-free cloth. Studio quality capacitor microphones are normally stand-mounted and protected with a pop screen or slip-on foam screen. Keep enough of these devices on hand for backup replacement.

Modern dynamic microphones are surprisingly rugged and can sustain multiple drops. Much the same can be said for small format capacitor models, but large format capacitors will usually end up with dented screens if dropped. In most variable pattern models, the dual diaphragm element is suspended on a small post. Some of the vintage models have been known to break at this point when dropped, so proceed with caution.

PERFORMANCE CHECKS

If there is any question about the frequency response of a microphone, it can be checked one of two ways. Figure 20–1 shows a method for comparing a questionable microphone with one known to be in excellent operating condition. First, use the reference microphone to arrive at a rough equalization for flat response as shown on the RTA. (This will facilitate seeing any clear differences between the response of the two microphones.) Then, using the same microphone location, substitute the questionable microphone and compare the two. A test such as this is not as rigorous as would be carried out in the manufacturer's laboratory, but it will enable you to identify a problem. You should also be aware that not all microphones of a given model are identical in frequency response. The published specifications will usually make this clear.

The test should be done in a fairly dead acoustical space and as isolated as possible from early reflections. Do not be overly concerned that

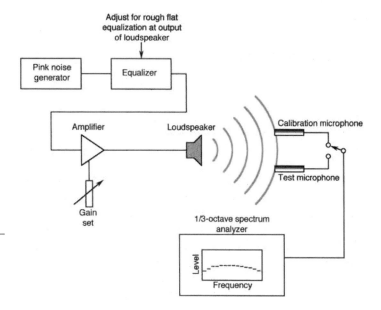

FIGURE 20-1

Comparing frequency response of two microphones using a real-time analyzer.

neither of the two response curves will look flat. What you are looking at here is a combination of both loudspeaker and microphone response, and your concern should be only with the differences between the two microphone curves.

An alternate method is shown in Figure 20–2. In this test the two microphones are placed as close together as possible, and a talker is placed about 1 m (40 in) away. Feed both microphones into a console and set both trims and faders to the same positions. Then, invert the polarity of one of the microphones. If the microphones are well matched the sound level produced by the control room monitors will drop considerably. You may need to make a fine adjustment to one of the trim controls in order to null out the response to the maximum degree. This test is not quite as rigorous at the first, but it will isolate any serious differences between the microphones.

You can get a rough idea of relative microphone sensitivities using the test setup shown in Figure 20–3. The test should be carried out in a studio and well away from any reflective boundaries. You will need a sound level meter (SLM) to set the level 1 m from the small loudspeaker to approximately 94 dB L_p as measured on the C or flat scale. In the laboratory a precision SLM would be used for this purpose, but our concern here is primarily with the relative sensitivities between pairs of microphones.

Both microphone and SLM should be located 1 m from the loudspeaker, and for most purposes the measurement should be made using

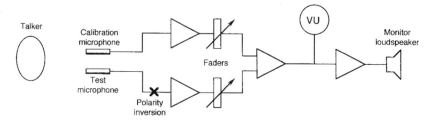

FIGURE 20-2

Comparing frequency response of two microphones by nulling their outputs.

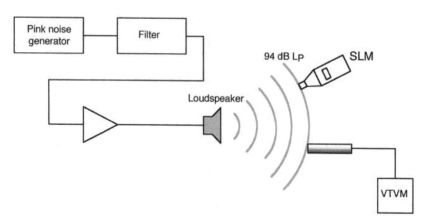

FIGURE 20-3

Comparing microphone sensitivity.

an octave-wide band of pink noise centered at 1 kHz. Rough sensitivity comparisons may be read directly from the meter settings on the console that are necessary to match their levels.

The relative self-noise levels of microphones may be compared by using an extension of the test we have just discussed. Adjust the two microphones so that their outputs, as measured in the control room and using a noise signal of 94 dB L_P, are equal. Now, without making any changes in the gain structure, turn off the noise source and move the microphones to a very distant and quiet location far away from the control room. A good place might be an isolation room or closet at the far end of the studio. The intent here is to prevent any audible acoustical feedback as you proceed with the rest of this test.

Progressively raise the gain of both microphones *equally*. Do this by rerouting each microphone through a second set of console inputs so that you will have enough reserve gain to raise the noise level to a point where you can clearly hear it. You will progressively have to lower the control room monitor loudspeakers as you do this, and it is also a good idea to roll off excess LF response of both microphones via the in-line equalizers. The key point here is to maintain exactly the same gain in both microphone channels. Be very careful! With all of this excess gain you could, with just a little carelessness, damage your monitor loudspeakers. You may need 50 to 60 dB of added gain in order to hear the self-noise levels clearly.

When carefully carried out this test will let you make subjective comparisons of the noise floors of a pair of microphones. The procedure is outlined in Figure 20–4.

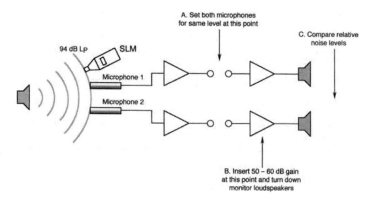

FIGURE 20-4 ————

Comparing microphone self-noise floors.

CHAPTER 21

CLASSIC MICROPHONES: THE AUTHOR'S VIEW

INTRODUCTION

This chapter was written as a result of numerous conversations with both microphone users and manufacturers following publication of the first edition of *The Microphone Book*. While the list reflects my personal view of the subject, it should come as no surprise that it contains virtually all of the celebrated models of the last 75 years. Most engineers would agree that the German and Austrian capacitor (condenser) models introduced into the United States shortly after the Second World War virtually reinvented both popular and classical recording in America, and those models form the core of this presentation. To these I have added the American contributions in dynamic and ribbon technology that became the mainstay of broadcasting and communications in this country, along with a group of fine omnidirectional capacitors that were also distinctively American.

There are several independent criteria I have set for inclusion in the list:

1. The microphone should be at least 30 years old and generally regarded as an exemplar in its field. Some models are still in demand, often at collectors' traditionally high prices.

2. Microphones that have been at the vanguard of a particular design or usage technology are also eligible, whether or not they have attained wide renown.

3. Microphones that at an earlier time attained a unique cult following. What I have in mind here are the aforementioned high-performance American capacitor omnis that flowered in the early 1950s as part of the burgeoning audiophile tape recording movement.

You might think that gathering photos and specifications of all of these entries would have been a simple task. Not so. While I want to thank many people for their permission to use photos of some of these models, the bulk of the graphics in this chapter come from older publications and manufacturers' original specification sheets, many of them requiring all of the touchup wizardry that modern computer graphics programs can provide. The presentation order is essentially chronological, with occasional grouping of similar models that may be slightly out of sequence.

Many of the entries show more than just a microphone. Response curves, circuit diagrams, and control functions are also presented where they underscore the performance of a given model.

This is as good a time as any to acknowledge the many avid authors, designers, and historians who have embraced the subject of microphones over the years. Their dedication to this fascinating field has kept a unique history alive and vital for many of us – Abaggnaro, 1979; Bauer, 1987; Knoppow, 1985; Paul, 1989; Sank, 1985; Webb, 1997; and Werner, 2002.

NEUMANN MODEL CMV3A

The stunning industrial design of this microphone takes its cue from the German Bauhaus movement, and its realization in fine metal casting and finishing is truly amazing for 1928 (Figure 21–1). While Wente's omnidirectional capacitor design (Western Electric 1917) predates Georg Neumann's design by a decade, we must remember that American capacitors were virtually all omnidirectional until well after the Second World War, whereas Neumann's design included bidirectional and cardioid elements based on the work of Braunmühl and Weber. Neumann's work in thin plastic material development and gold sputtering techniques pointed the way for low mass diaphragms with extended response and high sensitivity. You can hear a stereo pair of these microphones on a CD titled "Das Mikrofon" (Tacet 17) in a performance of excerpts from a Haydn string quartet. There is nothing "old" about the sound of these microphones.

Largely because of Neumann's pioneering work, the European broadcasting and recording communities adopted the capacitor microphone early on in their pursuit of sonic excellence – a state of affairs that would later position them favorably in the world market that developed after the war.

WESTERN ELECTRIC MODEL 618

When broadcasting was introduced in 1919, carbon button microphones were the mainstay of transmission in the studio. A few capacitors had also made the grade, but their expense and technical requirements had pretty much ruled them out for general use. The introduction of the Western Electric 618 dynamic microphone was a breakthrough in that is was rugged

FIGURE 21-1

Neumann Model CMV3A. (Photo courtesy of Neumann/USA.)

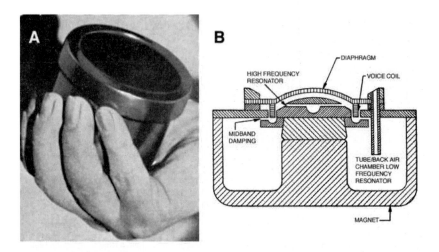

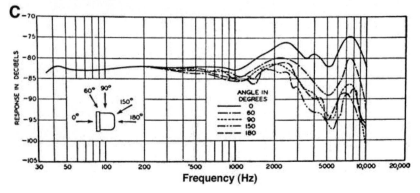

FIGURE 21-2

Western Electric Model 618
(A and C from Read, 1952;
B from Frayne and Wolfe,
1949.)

and, thanks to new cobalt-based magnet materials, had good sensitivity
and did not require a power supply. In essence it was the first dynamic
omnidirectional microphone, complete with all of the features that are
inherent in that design today. It had a midband-tuned highly damped
moving system complete with a resonance chamber for extended LF and
a small resonator behind the diaphragm for extended HF response.
Figure 21–2A shows the basic unit, and a cutaway view of the system is
shown at B. Typical on- and off-axis response curves are shown at C.

WESTERN ELECTRIC MODEL 640AA

Because of their extensive manufacturing of microphones for broadcasting, motion picture sound, and recording, Western Electric had a need
for a highly stable calibration microphone. The introduction of the
640AA in 1932 provided this reference unit, and it soon became a standard for all electroacoustical manufacturing in the US. Even after Western
Electric ceased microphone manufacturing in the early 1950s, it retained

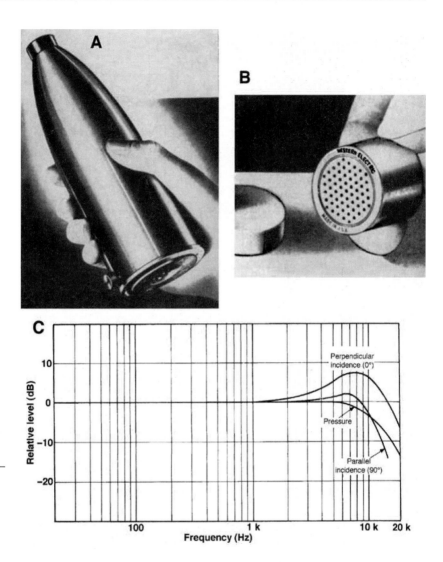

FIGURE 21-3

Western Electric Model 640AA. (A and B from Read, 1952; C from National Bureau of Standards, 1969.)

the manufacture of this model until the Danish Bruël & Kjær company had attained sufficient stature in the instrumentation microphone field to become the new standard. Figure 21–3A shows the 640AA unit mounted on the RA-1095 preamplifier, and a close-up detail of the capsule is shown at B. Typical response curves are shown at C.

WESTERN ELECTRIC MODEL 630

Introduced in the mid 1930s, the model 630, dubbed the "Eight Ball," improved on the performance of the 618. The new model was designed to be used with its principal axis facing upward so that it had uniform response over the entire 360° azimuthal plane. The spherical body design

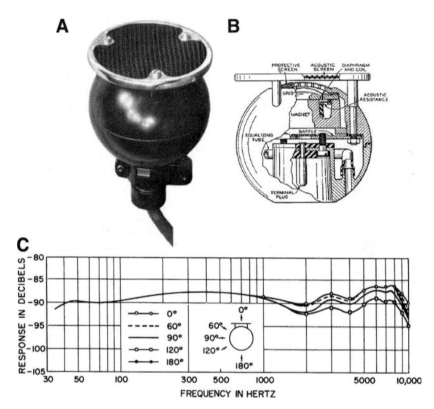

FIGURE 21-4

Western Electric Model
630. (A and C from
Tremaine, 1969; B from
Read, 1952.)

contributed to the overall smoothness of response. Note that the 90°
off-axis response is maintained within an envelope of ±3 dB from 50 Hz
to 10 kHz.

WESTERN ELECTRIC MODEL 639

This famous model from the late 1930s represents Western Electric's
only use of ribbon technology. In it, there are two elements: an omni
dynamic and a ribbon mounted directly above it. (See Figure 5–8 for
internal construction details.) While the RCA ribbons were corrugated
from top to bottom, the ribbon in the 639 was corrugated only at the top
and bottom, with the middle section crimped so that it moved as a unit.
Via external switching, the two acoustic elements could be operated
independently or combined in several ways to produce the family of first-
order cardioids. Figure 21–5A shows a view of the famous "bird cage"
design, typical on- and off-axis response curves are shown at B, C, and D
for three modes of response.

The model shown here bears the Altec brand name. By the mid
1950s, Western Electric ceased manufacturing commercial microphones,
giving that segment of their business to the Altec Corporation, which

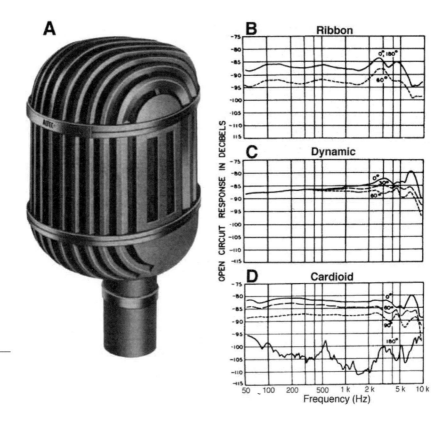

FIGURE 21-5

Western Electric
Model 639. (Data from
Tremaine, 1969.)

had earlier assumed the manufacture and sale of Western Electric loud-speakers and amplifiers in the late 1930s.

RCA MODEL 44

The original A-version of this most famous of all ribbon microphones dates from 1931, with the BX-version following in the mid 1930s. Few microphones are as distinctive in overall shape, and its "box-kite" shape has been copied many times. Olson's dedication to the engineering simplicity of the ribbon was a hallmark of his illustrious 50-year career at RCA Laboratories, and a quick survey of Olson's writings will underscore the variety of concepts, models, and design challenges that he solved using only ribbons. The microphone became a broadcast favorite during the 1930s, lasting well into the 1960s. It was also a staple in the recording studio from the 1930s onward and more than held its own against the onslaught of German and Austrian capacitor models as the LP era got under way.

You will find 44s, lovingly maintained, in large recording complexes everywhere, where they are used primarily for close-in brass pickup in large studio orchestras and big band jazz groups. Many engineers feel that the directivity and proximity effects of the 44 provide a warmth and mellowness on loud, hard-driven brass passages that can't be easily attained using capacitors, no matter how many equalizers you may throw at the problem!

The rugged look of these microphones belies their essential fragility. The ribbon elements themselves are low-frequency tuned and as such can be deformed under conditions of mechanical shock. The perforated protective metal screening is also very prone to denting if the microphone is dropped. A front view of the microphone is shown in Figure 21–6A, and typical on-axis response is shown at B. RCA ceased manufacture of this model in the early 1970s.

A

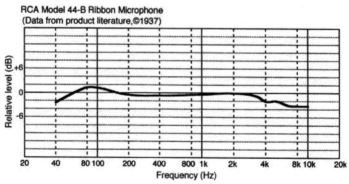

B

RCA Model 44-B Ribbon Microphone
(Data from product literature,©1937)

FIGURE 21-6

RCA Model 44-BX.
(A from Tremaine, 1969; B
from RCA specification
sheet.)

RCA MODEL 77

Virtually everything we have said about the RCA 44 applies as well to the 77. The microphone was designed by Olson's group in the 1930s as a unidirectional (cardioid) model for those applications where the figure-8 response of the 44 was not appropriate. The microphone provided a far more varied design test bed than the 44, and the 77 ultimately went through 7 iterations (Webb, 1997). A rear view showing the pattern adjusting control is shown in Figure 21–7A, and three switchable polar pattern families are shown at B, C, and D.

Earlier models of the 77 used the technique shown in Figure 5–9 to attain cardioid response. One portion of the ribbon was exposed on both

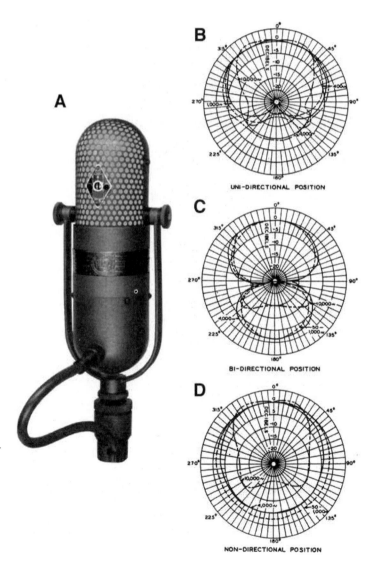

FIGURE 21-7 ————

RCA Model 77-D: back view of microphone showing pattern adjustment screw control (A); polar response (B, C, and D) (Data from Tremaine, 1969.)

sides and responded with a figure-8 pattern. The other half of the ribbon was shrouded on one side via a damping tube and responded as a pressure element. The two combined to produce a cardioid pattern. Later models used a variable approach as shown in Figure 5–15A. Here, the shroud extended over the entire back side of the ribbon, and an aperture, adjustable by a movable vane, provided the amount of gradient component required to attain a desired first-order pattern. As we see at B, C, and D, the patterns were fairly accurate up to about 4 kHz; above that frequency the back-side pattern response was only approximate.

RCA MODEL MI-10001

This 1947 model was based on the later 77 designs. It had a nonadjustable shroud tailored to produce a nominally flat cardioid pattern aimed downward 45°. It was intended for boom use on motion picture

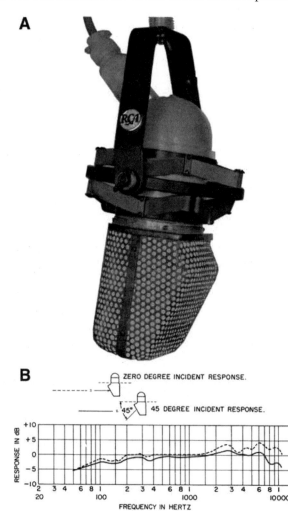

FIGURE 21–8

RCA Model MI-10001: view of microphone (A); response curves (B) (Data from Tremaine, 1969.)

FIGURE 21-9 ————

Neumann Model U47.
(Photo courtesy of
Neumann/USA.)

sound stages and was produced in limited quantities. It became the first microphone to be supported in field literature by a comprehensive three-dimensional family of polar response curves, of which only two are shown in Figure 21–8.

NEUMANN MODEL U47

This multiple pattern microphone was introduced to the US market in 1948 under the trade name of Telefunken, a German distribution company, and gained a high reputation for excellence in both popular and classical recording. Its introduction coincided pretty much with the development and advent of the LP record, and it was highly touted (often as a choice item for LP cover art) for its extended high frequency response. At a price tag of about $500 (in 1948 a great deal of money), it soon became a favorite of the recording elite.

The microphone, which is shown in Figure 21–9, provided both cardioid and omnidirectional patterns. A close relative, the U-48, was introduced in the mid-1950s and offered cardioid and figure-8 patterns. By the late 1950s Neumann began marketing its products directly in the US, and the Telefunken brand name was replaced by Neumann. It remains one of the most sought-after classic European capacitor models.

NEUMANN MODELS M49/50

These two models are completely different but share the same outer packaging envelope. The M49, introduced in 1949, was the first capacitor microphone to offer remote pattern switching via a control on the power supply, as can be seen at B in Figure 21–10. The M50, which is described in detail in Chapter 3 under On-axis Versus Random Incidence Response, was designed to solve orchestral pickup problems when positioned in the transition zone between direct and reverberant fields of the ensemble. Its capsule is mounted on a plastic sphere and behaves as a pressure element up to about 2.5 kHz. Above that point the response becomes gradually more directional and rises in output, attaining a +6 dB shelf in response relative to mid and lower frequencies. The intention was to maintain a certain amount of sonic intimacy when recording at normal distances in spaces with moderate reverberation.

The M50 became the microphone of choice for both British Decca and EMI classical engineers, and the so-called Decca tree, plus outriggers, made use of five M50s. Today, it is a mainstay of classical orchestral recording in large spaces, and it is also widely used for pickup of stereo ambience in large studios and soundstages.

AKG ACOUSTICS MODEL C-12

AKG was founded in Vienna in 1947 and quickly became a major player in the field of recording and broadcasting microphones. Details of the

A

B

FIGURE 21-10 ————

Neumann Models M49/50.
(Photos courtesy of
Neumann/USA.)

C-12 capacitor model, which was introduced in 1953, are shown in
Figure 21–11. Electrical pattern switching was facilitated by the unique
dual backplate capsule design used by AKG. Note the pattern switching
unit, which is in line with the microphone and the power supply (detail
shown at C). Switching was noiseless and could be carried out during
recording. The C-12 is highly sought after today for studio work, most
notably for pop vocals.

TELEFUNKEN ELAM 251

When Neumann took over its own foreign distribution, Telefunken
turned to AKG for replacement models and in 1959 contracted for what
became the ELAM 251. The basic elements in this model were the same

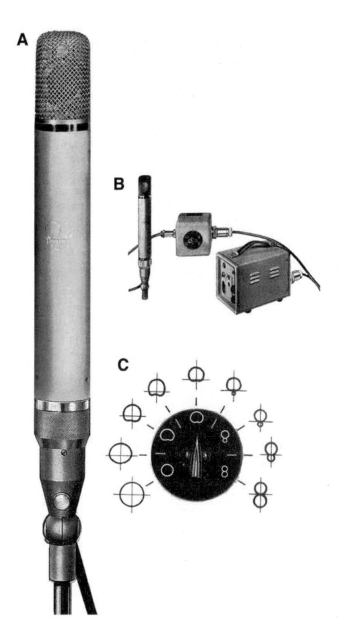

FIGURE 21-11 ———

AKG Model C12. (Data
courtesy of AKG
Acoustics.)

as in the C-12; however, the in-line pattern changing box was incorporated as a three-way switch on the microphone body. In the overall
redesign process the microphone body was made somewhat larger. The
ELAM 251 remains today one of the most prized, and expensive, vintage
tube microphones items in any studio.

Those readers interested in more technical details of the C12 and
ELAM 251 designs are referred to Paul (1989) and Webb (1997).

SHURE MODEL M-55 UNIDYNE

Introduced in the late 1930s, the Shure M-55 Unidyne was the first dynamic microphone with a cardioid pattern. It became the industry prototype for all dynamic vocal microphones both in the US and in Europe. It was invented by the ever-resourceful Benjamin Bauer, who reasoned that a good cardioid pattern should exist somewhere between omni and dipole. His simple solution was to introduce a two-element acoustical phase shift network instead of a fixed delay path in one branch of a basic dipole. The time constant of the phase shift network provided excellent front-back rejection over a broad portion of the midrange, while forward directionality provided the necessary front-back discrimination at higher frequencies. The M-55 is shown in Figure 21–13A, and on-axis response is shown at B.

The M-55 has been in the Shure catalog in one form or another since its inception in the late 1930s. It is truly one of the icons of the industry and has been documented in press and news activities over the years. Who can forget those photos of Elvis Presley and President Kennedy with the ubiquitous M-55?

FIGURE 21-12

Telefunken Model ELAM 251. (Data from early company advertising.)

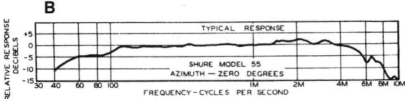

FIGURE 21-13

Shure Model M-55: photo (A); typical on-axis response (B). (Data courtesy of Shure Inc.)

ELECTRO-VOICE MODEL 666 VARIABLE-D®

Wiggins (1954) designed the Variable-DD® dynamic cardioid at Electro-Voice. Up to that time, all dynamic cardioid designs had a single rear opening and a mass-controlled diaphragm in order to maintain flat, extended LF response. The penalty paid here is handling noise and susceptibility to mechanical shock. By stiffening the diaphragm and providing three rear openings (low, medium, and high frequency), each of a different length, the net forces on the diaphragm remain as before, and the response is effectively flat. (Design details are given in Chapter 5 under The Electro-Voice Variable-D® Dynamic Microphone.)

The model 666 is shown in Figure 21–14A. That microphone, in combination with the equalizer shown at B, comprise the Model 667 system. In addition to normal speech applications, Electro-Voice recommended the 666 for more distant sound stage applications, inasmuch as the microphone had good noise immunity. The electronics package

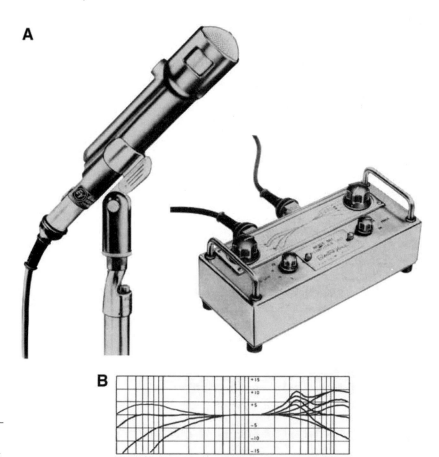

FIGURE 21-14 ———

Electro-Voice Model 666/667. (Data courtesy of Electro-Voice.)

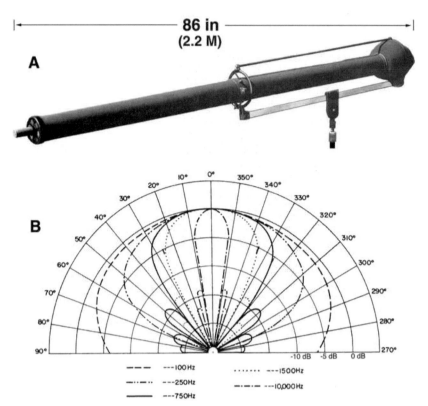

FIGURE 21–15

Electro-Voice Model 643
rifle microphone. (Data
from Tremaine, 1969 and
company advertising.)

provided added gain and a family of EQ curves to correct for various distance effects. Details of the EQ contours are shown at C.

ELECTRO-VOICE MODEL 643 RIFLE MICROPHONE

Over the years, rifle microphones have rarely exceeded about 0.5 m (20 in) in length, but in the late 1950s Electro-Voice designed the largest commercial model ever, the 643, which measured 2.2 m (86 in). The model was designed for coverage of sports events and other large venue activities in an era long before wireless microphones were commonplace. The microphone is shown in Figure 21–15A, and polar response curves are shown at B. While the 643 did a remarkable job at moderate distances, ambient noise, much of it at middle and low frequencies, took its toll on intelligibility. Additionally, the microphone was expensive and clumsy to set up and operate. With the coming of wireless microphones the distant coverage challenges could be far more easily met.

AKG ACOUSTICS MODEL D-12

The D-12 dynamic cardioid microphone was introduced in 1952 and with some modifications remained in the AKG catalog until 1985. It was notable

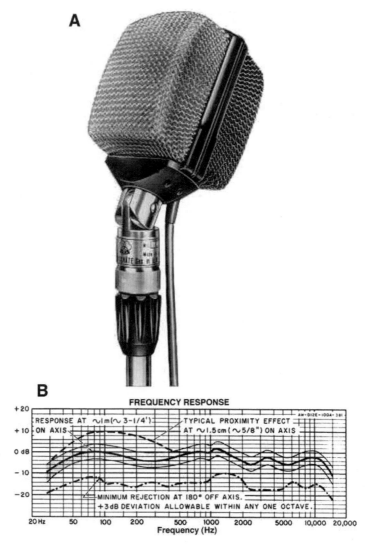

FIGURE 21-16

AKG Acoustics Model D-12: photo of microphone (A); various response curves (B). (Data courtesy of AKG Acoustics.)

in studio work for its abilities to handle high levels and became a favorite, both in Europe and the US, for kick drum pickup. The design of the microphone included a very linear moving system and a LF resonance chamber which extended the response down to about 30 Hz. Figure 21–16A shows a photo of the model, and frequency response curves are shown at B. You will find many D-12s still in use, and the model is one of very few single diaphragm dynamic cardioids to have attained such vintage status.

SONY MODEL C-37A CAPACITOR MICROPHONE

The C-37A quickly became a lower cost alternative to the Telefunken U47 when it was introduced during the mid 1950s, and it is easy to see why. As is clear from Figure 21–17B, the response was remarkably flat to 20 kHz, lacking the rise in the 10 kHz range which characterized many of the European dual diaphragm capacitors. Notable also is the integrity of the cardioid pattern, which retains its ±90° target response of −6 dB remarkably well to just below 10 kHz. By any measure this is excellent response from a nominal 25 mm (1 in) diameter diaphragm. The only negative design comment that can be made is the susceptibility of the perforated metal screen to show dents – much the same way the RCA ribbons do.

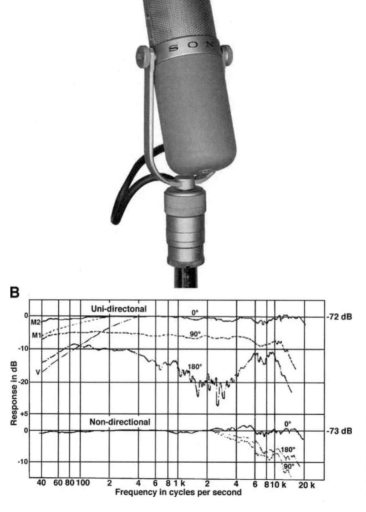

FIGURE 21-17

Sony Model C-37A dual pattern capacitor. (Photo courtesy of Eric Weber; data at B from Sony specification sheet.)

THE FIRST STEREO MICROPHONES

The promise of stereo in the mid 1950s gave rise to a number of capacitor microphone models in which rotatable, variable pattern elements were stacked vertically. Notable models here were made by Neumann, AKG Acoustics, and Schoeps, and they were all quite similar in overall design. Only the capsule at the end of the stem was rotatable, but both capsules had variable patterns. All were tube types, but the designs changed rapidly with the coming of solid state electronics and phantom powering during the 1960s. The Neumann and AKG models used dual diaphragm capsules while the Schoeps model used their hallmark single diaphragm capsule with acousto-mechanical pattern changing.

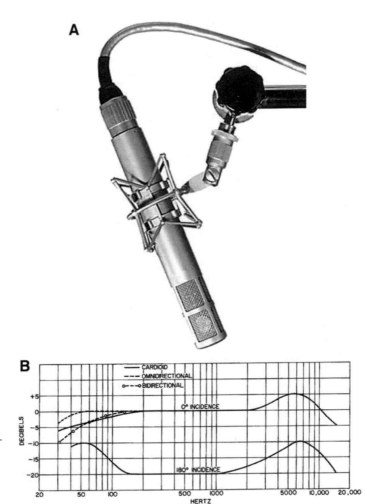

FIGURE 21-18

Neumann Model SM2 stereo microphone. (Data courtesy of Neumann/USA.)

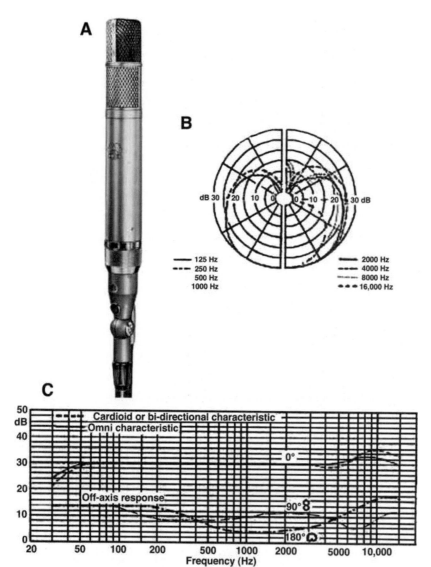

FIGURE 21-19

AKG Acoustics Model C-24 stereo microphone. (Data courtesy of AKG Acoustics.)

NOTABLE AMERICAN CAPACITOR OMNIS

As the 1950s got under way it seemed that high-end microphone technology was rapidly becoming a European specialty. In all other respects, including basic recording technology, loudspeakers, and the electronic chain, American companies were very much in the running – and in fact had led in the early development of the post-war art. But in the microphone area, American companies had invested heavily in dynamic and ribbon technology while the recording industry had firmly endorsed the newer capacitor microphones as the leading edge in design.

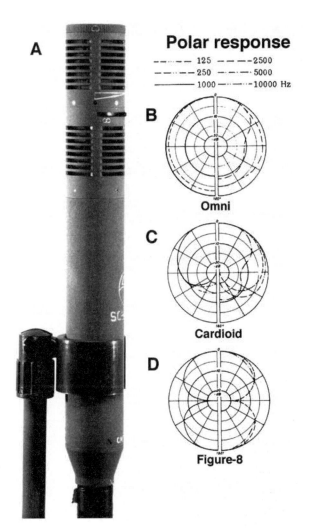

Polar response
    ——··—— 125 ————2500
    ——··—— 250 ——·—·· 5000
    ———— 1000 ——··——10000 Hz

B

Omni

C

Cardioid

D

Figure-8

FIGURE 21-20 ————

Schoeps Model CMTS-501 stereo microphone. (Data courtesy of Schoeps GmbH.)

A handful of American companies, some of them quite small, responded to this challenge. Noting that omni capacitors were relatively straightforward in concept, they decided to make their individual contributions in this direction. Those companies were: Altec, the most prominent manufacturer of systems for motion pictures and sound reinforcement; Stephens Tru-Sonic, a manufacturer of high-end loudspeakers; Capps and Company, a manufacturer of disc recording styli; and a small California company, Stanford-Omega.

The Altec M21 system is shown in Figure 21–21A. The basic design was begun in the late 1940s, and a number of capsule designs were experimented with. The version shown here was unique in that sound reached the diaphragm indirectly through a circular opening around the rim, which can be seen in the cutaway view at B. The diaphragm itself was a piece of optical quality glass ground to a thickness of approximately 0.013–0.02 in,

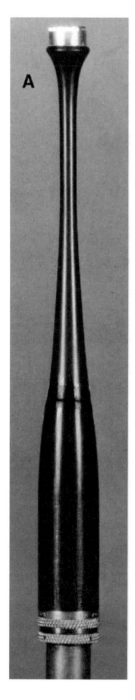

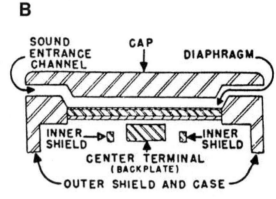

FIGURE 21-21 ————

Altec Model M21 capacitor
microphone: photo of 150A
base with M21B capsule
(A); cutaway view of
capsule (B). (Photo courtesy
of William Hayes and
Micromike Labs.)

depending on the desired sensitivity, and gold plated on one side. No
tensioning of the diaphragm was necessary inasmuch as the stiffness of
the thin glass was sufficient to maintain a high resonance frequency.
A more conventional capsule, the 11B was also used with the same base.

The Stephens Tru-Sonic company was founded in Los Angeles in the 1940s by Robert Stephens, who had worked with Lansing, Shearer, and Hilliard on the academy award winning MGM loudspeaker system for motion pictures. Like Lansing, he absorbed the manufacturing philosophies and techniques that had been pioneered by Western Electric, and his company was regarded during the 1950s in the same professional class as Altec and JBL. In the early 1950s he introduced the first RF microphone intended for recording and reinforcement activities, the model C2-OD4. Figure 21–22A shows the C2 capsule assembly mounted on a narrow stem, which was connected to the OD4 oscillator-demodulator through a coaxial cable of any length – as long as the overall cable length (transmission line) was a multiple of 37.5 inches – a requirement for efficient signal transfer from the capsule to the demodulator. The schematic of the oscillator-demodulator is shown at B, and it can be seen that the capsule portion contained only the half-inch (12 mm) diaphragm closely coupled to an oscillating tank circuit. The system operated on the FM

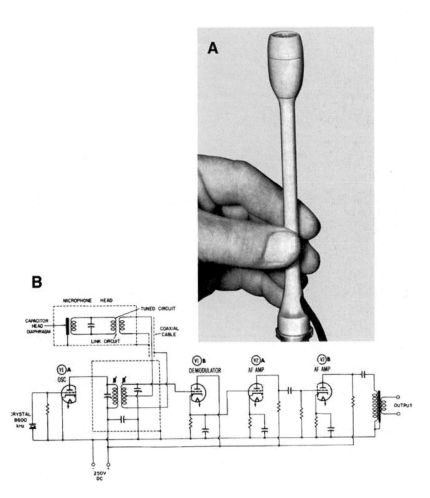

FIGURE 21-22

Stephens Model C2-OD4 RF microphone. (Photo A from Read, 1952; data at B from Tremaine, 1969.)

principle in the range of 9 MHz, and signal demodulation took place in the associated circuitry.

The system had excellent performance characteristics but exhibited certain reliability problems due to shifts in the operating characteristics of the vacuum tubes in the electronics unit. The microphone was a favorite of such notables as Ewing Nunn, producer of Audiophile recordings, and Paul Klipsch, noted loudspeaker manufacturer and recording enthusiast. Perhaps because of the long term stability problem, relatively few of these microphones were built.

Frank Capps had worked for Edison in his phonograph division, and when that company closed in 1929 Capps founded a company whose

FIGURE 21-23

Capps Model CM2250 microphone. (Data from company advertising.)

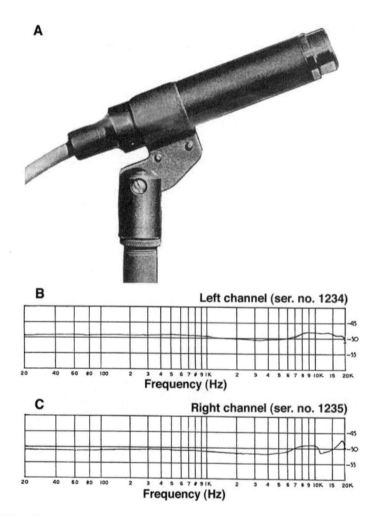

FIGURE 21-24

Stanford-Omega Condenser Microphone: view of microphone (A); frequency response curves for a stereo pair (B and C). (Data provided by Lowell Cross.)

single product was cutting styli for disc recording. In the early 1950s the company developed the model CM2250 capacitor omni shown in Figure 21–23. The design was elegant and the tapered capsule assembly was easily spotted at some distance. Emory Cook, of Cook Laboratories, used a pair of these microphones in virtually all of his early audiophile and ethnic music recordings.

Marketed under the name Stanford-Omega, the Thompson Omega company in southern California manufactured the microphone shown in Figure 21–24A. The microphone carried no model number and was known only as the Stanford-Omega Condenser Microphone. The manufacturer did something that very few companies do today – they provided actual calibration curves on each microphone. Curves for a typical stereo pair are shown at B and C.

BANG&OLUFSEN MODEL BM-3

The Danish Bang&Olufsen company is primarily noted for high performance home high fidelity equipment elegantly representative of modern Scandinavian industrial design. As far back as the early 1950s the company was experimenting with stereo, and a variety of ribbon microphones, both mono and stereo, were built by the company. Compared with the relatively large ribbons of the mid 1950s, the B&O models were fairly small, and response beyond 10 kHz was maintained. The microphone was made famous by the many articles written by Erik Madsen (1957) in both the technical and popular press in which a spaced pair of these microphones were deployed with a directional baffle (see Figure 12–9A).

Figure 21–25A shows a view of the BM-3 model; response curves, with and without the protective mesh grille, are shown at B. Note the smooth response and HF extension beyond 10 kHz. The exposed flanges

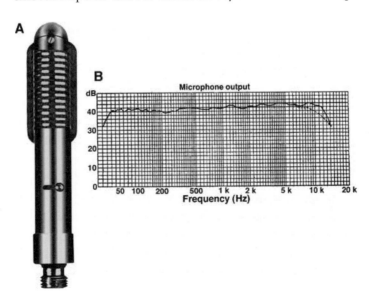

FIGURE 21-25

B&O Model BM-3 ribbon microphone: Photo (A); frequency response (B). (Data from B&O specification sheet.)

on each side of the grille structure are actually magnetic return paths around the ribbon.

THE RISE OF SMALL FORMAT CAPACITOR MICROPHONES

The first capacitor microphones for studio applications were fairly large models, most of them with switchable pickup patterns. During the early to mid 1950s three companies, AKG, Schoeps, and Neumann introduced smaller format designs with capsules in the diameter range 15–18 mm

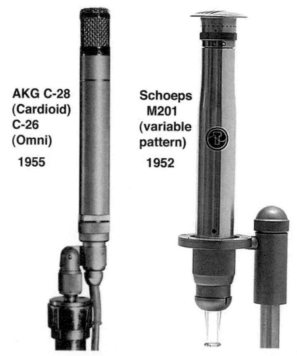

AKG C-28 (Cardioid) C-26 (Omni) 1955

Schoeps M201 (variable pattern) 1952

Neumann KM 53 (Omni); KM 54 (Cardioid) 1953

FIGURE 21-26 ───────

The earliest small format capacitor microphones from AKG, Schoeps and Neumann. (Photos courtesy of the manufacturers.)

(0.6–0.7 in). Many of these had families of replaceable capsules that could operate on a common preamplifier. Figure 21–26 shows three such early models, one from each company, that typified this design approach. With the advent of phantom powering in the 1960s these designs proliferated and, with their lower prices, eventually dominated the professional market, at least in terms of numbers.

SENNHEISER MODEL MKH 404 RF MICROPHONE

Taking advantage of solid state electronic components and remote powering, Sennheiser Electronics introduced its RF microphones in the early 1960s. The designs have been refined over the years, and today they are the flagship products of the company, exhibiting performance characteristics that are at the leading edge of the art. The MKH 404, shown at Figure 21–27A, was the first solid state cardioid microphone

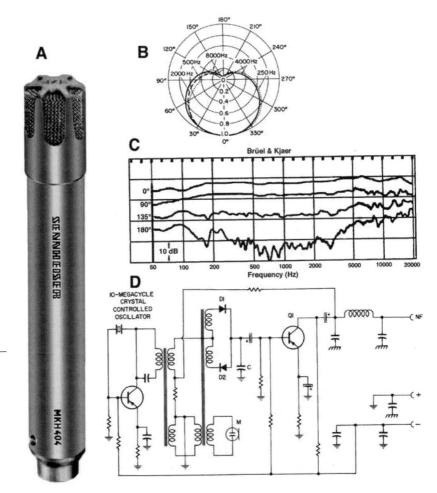

FIGURE 21-27

Sennheiser MKH 404 RF microphone: photo of microphone (A); polar response (B); off-axis response curves (C); circuit diagram (D). (Data courtesy of Sennheiser Electronics and Tremaine, 1969.)

that operated on the RF principle. The design had been preceded by the omnidirectional model MKH 104. Performance data is shown at B and C, and a schematic drawing of the electronics is shown at D. The overall uniformity of response is notable, but the polar response is exemplary, even by today's standards. In the earliest days of these models, phantom powering was not yet in general use; Sennheiser manufactured an in-line battery module that provided 9 Vdc powering for these microphones.

SENNHEISER MODEL MD 421 DYNAMIC MICROPHONE

The famous 421 was introduced in 1960 and has been in the Sennheiser catalog ever since. While it was intended as a general-purpose dynamic cardioid, its robustness was evident from the start. It eventually gravitated, along with the AKG D-12, into its present role as a kick drum microphone par excellence. Today it is the most sought-after dynamic microphone for that purpose.

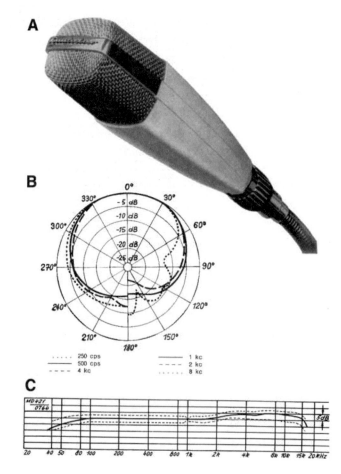

FIGURE 21-28

Sennheiser Model MD 421 dynamic. (Data courtesy of Sennheiser Electronics.)

CROWN INTERNATIONAL MODEL PZM-30
BOUNDARY LAYER MICROPHONE

Dating from the early 1970s, this is the "youngest" microphone presented in this survey. The Crown PZM pressure zone microphones differ from other omni boundary microphones in that they provide a short indirect path to the microphone capsule, thus ensuring that the amplitude pickup will not be materially influenced by the bearing angle of the sound source. The microphone is shown in Figure 21–29A, and details of the actual path to the capsule are shown at B. The PZM-30 has been in the Crown catalog since it was introduced and remains the classic reference for boundary layer recording.

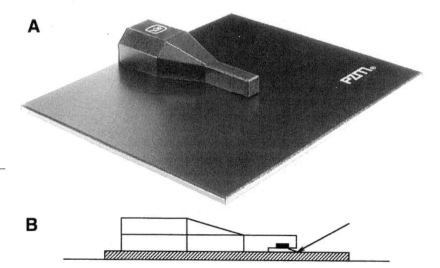

FIGURE 21-29 ———

Crown International PZM-30 boundary layer microphone: photo (A); detail of sound pickup (B). (Photo courtesy of Crown International.)

REFERENCES AND BIBLIOGRAPHY

GENERAL AND HISTORICAL

L. Abaggnaro, ed., *AES Anthology, Microphones*. New York: Audio Engineering Society, 1979.

Anon., "The Telephone at the Paris Opera." *Scientific American*, 31 December 1881 (reprinted in *Journal of the Audio Engineering Society* 29, no. 5 (May 1981).

M. Barron, "The Subjective Effects of First Reflections in Concert Halls – the Need for Lateral Reflections." *Journal of Sound and Vibration* 15 (1971), pp. 475–494.

B. Bauer, "Uniphase Unidirectional Microphone." *Journal of the Acoustical Society of America* 13 (1941), p. 41.

B. Bauer, "A Century of Microphones." *Proceedings, IRE* (1962), pp. 719–729 (Also, *Journal of the Audio Engineering Society* 35, no. 4 (1987).

A. Benade, *Fundamentals of Musical Acoustics*. New York: Oxford University Press (1976).

L. Beranek, *Acoustics*. New York: J. Wiley, 1954.

L. Beranek, *Concert and Opera Halls: How they Sound*. Acoustical Society of America, New York (1996).

W. Bevan et al., "Design of a Studio Quality Condenser Microphone Using Electret Technology," *Journal of the Audio Engineering Society* 26, no. 12 (1978). (Included in AES Microphone Anthology.)

J. Blauert, *Spatial Hearing*. Cambridge, MA: MIT Press, 1983, pp. 209–210.

G. Boré, *Microphones*. Neumann USA, Sennheiser Electronics, 6 Vista Drive, Old Lyme, CT, 06371 (1989).

M. Brandstein and D. Ward (eds), *Microphone Arrays*, New York: Springer, 2001.

H. Braünmuhl and W. Weber, "Kapacitive Richtmikrophon," *Hochfrequenztechnic und Elektroakustic* 46 (1935), p. 187.

D. Cooper and T. Shiga, "Discrete-Matrix Multichannel Sound," *Journal of the Audio Engineering Society* 20, no. 5 (June 1972).

J. Cooper, *Building a Recording Studio*, 5th edn. Synergy Group, 23930 Craftsman Road, Calabasas, CA 91302 (1996).

S. Dove, "Consoles and Systems," Chapter 22 in *Handbook for Sound System Engineers*. Indianapolis: Sams & Co., 1987.

J. Eargle, *Microphone Handbook*. Commack, NY: Elar Publishing, 1981.

J. Eargle, *Electroacoustical Reference Data*. New York: Van Nostrand Reinhold, 1994.

J. Eargle, *Music, Sound & Technology*. New York: Van Nostrand Reinhold, 1995.

J. Frayne and H. Wolfe, *Sound Recording*. New York: John Wiley & Sons, 1949.

M. Gayford, ed., *Microphone Engineering Handbook*. London: Focal Press, 1994.

R. Glover, "A Review of Cardioid Type Unidirectional Microphones," *Journal of the Acoustical Society of America* 11 (1940), pp. 296–302.

M. Hibbing, "High-quality RF Condenser Microphones," Chapter 4 in *Microphone Engineering Handbook*, ed. M. Gayford. London: Focal Press, 1994.

F. Hunt, *Electroacoustics*. New York: Acoustical Society of America, 1982.

D. Josephson, "Progress in Microphone Characterization – SC-04–04," presented at the 103rd Audio Engineering Society Convention, New York, September 1997, preprint number 4618.

F. Khalil et al., "Microphone Array for Sound Pickup in Teleconference Systems," *Journal of the Audio Engineering Society* 42, no. 9 (1994).

L. Kinsler et al., *Fundamentals of Acoustics*. New York: J. Wiley & Sons, 1982.

R. Knoppow, "A Bibliography of the Relevant Literature on the Subject of Microphones," *Journal of the Audio Engineering Society* 33, no. 7/8 (1985).

Y. Mahieux et al., "A Microphone Array for Multimedia Workstations," *Journal of the Audio Engineering Society* 44, no. 5 (1996).

J. Meyer, *Acoustics and the Performance of Music*, translated by Bowsher and Westphal. Frankfurt: Verlag Das Musikinstrument, 1978.

J. Monforte, "Neumann Solution-D Microphone," *Mix Magazine* (October 2001).

H. Nomura and H. Miyata, "Microphone Arrays for Improving Speech Intelligibility in a Reverberant or Noisy Space," *Journal of the Audio Engineering Society* 41, no. 10 (1993).

H. Olson, "The Ribbon Microphone," *Journal of the Society of Motion Picture Engineers* 16 (1931), p. 695.

H. Olson, "Line Microphones," *Proceedings of the IRE*, vol. 27 (July 1939).

H. Olson, *Acoustical Engineering*, New York: D. Van Nostrand and Company, 1957. Reprinted by Professional Audio Journals, 1991, PO Box 31718, Philadelphia, PA 19147–7718.

H. Olson, "Ribbon Velocity Microphones," *Journal of the Audio Engineering Society* 18, no. 3 (June 1970). (Included in *AES Anthology, Microphones*.)

H. Olson, *Modern Sound Reproduction*, New York: Van Nostrand Reinhold, 1972.

P. Parkin, "Assisted Resonance," *Auditorium Acoustics*. London: Applied Science Publishers, 1975. pp. 169–179.

S. Paul, "Vintage Microphones," parts 1, 2, and 3, *Mix Magazine* (Oct, Nov, and Dec 1989).

C. Perkins, "Microphone Preamplifiers – A Primer," *Sound & Video Contractor* 12, no. 2 (February 1994).

S. Peus, "Measurements on Studio Microphones," presented at the 103rd Audio Engineering Society Convention, New York, September 1997, preprint number 4617.

S. Peus and O. Kern, "TLM170 Design," *Studio Sound* 28, no. 3 (March 1986).

J. Pierce, *The Science of Musical Sound*. New York: Scientific American Books, 1983.

O. Read, *The Recording and Reproduction of Sound*. Indianapolis: H. Sams, 1952.

A. Robertson, *Microphones*. New York: Hayden Publishing, 1963.

J. Sank, "Microphones," *Journal of the Audio Engineering Society* 33, no. 7/8 (July/August 1985).

G. Sessler and J. West, "Condenser Microphone with Electret Foil," *Journal of the Audio Engineering Society* 12, no. 2 (April 1964).

D. Shorter and H. Harwood, "The Design of a Ribbon Type Pressure-Gradient Microphone for Broadcast Transmission," Research Department, BBC Engineering Division, BBC Publications, London (December 1955).

H. Souther, "An Adventure in Microphone Design," *Journal of the Audio Engineering Society* 1, no. 2 (1953). (Included in *AES Microphone Anthology*.)

J. Steinberg and W. Snow, "Auditory Perspective – Physical Factors," *Electrical Engineering* 53, no. 1 (1934), pp. 12–15 reprinted in *AES Anthology of Stereophonic Techniques* (1986).

R. Streicher and W. Dooley, "The Bidirectional Microphone: A Forgotten Patriarch," AES 113rd Convention (Los Angeles 5–8 October 2002). Preprint number 5646.

H. Tremaine, *Audio Cyclopedia*. 2nd edn. Indianapolis: H. Sams, 1969.

M. van der Wal et al., "Design of Logarithmically Spaced Constant-Directivity Transducer Arrays," *Journal of the Audio Engineering Society* 44, no. 6 (1996).

Various, *Condenser Microphones and Microphone Preamplifiers: Theory and Application Handbook*, published by Brüel & Kjær, 185 Forest Street, Marlborough, MA 01752 (reprinted May 1977).

Various, *A History of Engineering & Science in the Bell System, The Early Years (1875–1925)*, Bell Laboratories (1975).

Various, "Shields and Grounds," *Journal of the Audio Engineering Society* 43, no. 6 (June 1995).

J. Webb, "Twelve Microphones that Changed History," *Mix Magazine* 21, no. 10 (October 1997).

B. Weingartner, "Two-Way Cardioid Microphone," *Journal of the Audio Engineering Society* 14, no. 3 (July 1966).

E. Werner, "Selected Highlights of Microphone History," AES 112th Convention (Munich, 10–13 May 2002). Preprint number 5607.

R. Werner, "On Electrical Loading of Microphones," *Journal of the Audio Engineering Society* 3, no. 4 (October 1955).

A. Wiggins, "Unidirectional Microphone Utilizing a Variable Distance between Front and Back of the Diaphragm," *Journal of the Acoustical Society of America* 26 (Sept 1954), pp. 687–602.

G. Wong and T. Embleton, eds, *AIP Handbook of Condenser Microphones*. New York: American Institute of Physics, 1995.

A. Zukerwar, "Principles of Operation of Condenser Microphones," Chapter 3 in *AIP Handbook of Condenser Microphones*, Wong, G. and Embleton, T., eds. New York: AIP Press, 1994.

STEREO AND MULTICHANNEL RECORDING

B. Bartlett, *Stereo Microphone Techniques*. Boston: Focal Press, 1991.

A. Benade, "From Instrument to Ear in a Room: Direct or via Recording," *Journal of the Audio Engineering Society* 33, no. 4 (1985).

A. D. Blumlein, "British Patent Specification 394,325 (Directional Effect in Sound Systems)," *Journal of the Audio Engineering Society* 6, (reprinted 1958), pp. 91–98.

J. Borwick, *Microphones: Technology and Technique*. London: Focal Press, 1990.

J. Bruck, "The KFM 360 Surround Sound – A Purist Approach," presented at the 103rd Audio Engineering Society Convention, New York, November 1997. Preprint number 4637.

R. Caplain, *Techniques de Prise de Son*, (in French). Paris: Editions Techniques et Scientifiques Françaises, 1980.

C. Ceoen, "Comparative Stereophonic Listening Tests," *Journal of the Audio Engineering Society* 20. no. 1 (1970).

H. Clark et al., "The 'Stereosonic' recording and reproduction system," *Journal of the Audio Engineering Society* 6, no. 2 (April 1958).

E. Cohen and J. Eargle, "Audio in a 5.1 Channel Environment," presented at the 99th Audio Engineering Society Convention, New York, October 1995. Preprint number 4071.

D. Cooper and J. Bauck, "Prospects for Transaural Recording, *Journal of the Audio Engineering Society* 37, no. 1/2 (1989).

J. Culshaw, *Ring Resounding*. New York: Viking, 1957.

J. Culshaw, *Putting the Record Straight*. New York: Viking, 1981.

N. Del Mar, *Anatomy of the Orchestra*, Berkeley and Los Angeles: University of California Press, 1983.

M. Dickreiter, *Tonemeister Technology* (translated by S. Temmer) New York: Temmer Enterprises, 1989.

W. Dooley, *Users Guide, Coles 4038 Studio Ribbon Microphone*, Pasadena, CA: Audio Engineering Associates, 1997.

W. Dooley and R. Streicher, "MS. Stereo: A Powerful Technique for Working in Stereo," *Journal of the Audio Engineering Society* 30 (1982), pp. 707–717.

J. Eargle, *Handbook of Recording Engineering*. Boston: Kluwer Academic Publishers, 2003.

N. V. Franssen, *Stereophony*. Eindhoven: Philips Technical Bibliography, 1963.

F. Gaisberg, *The Music Goes Round*. New York: Macmillan, 1942.

R. Gelatt, *The Fabulous Phonograph*. New York: Lippincott, 1955.

H. Gerlach, "Stereo Sound Recording with Shotgun Microphones," Product usage bulletin, Sennheiser Electronic Corporation, 6 Vista Drive, Old Lyme, CT (1989).

M. Gerzon, "Periphony: With-Height Sound Reproduction," *Journal of the Audio Engineering Society* 21, no. 1 (1973).

M. Gerzon, "Optimum Reproduction Matrices for Multichannel Stereo," *Journal of the Audio Engineering Society* 40, no. 7/8 (July/August 1992).

M. Hibbing, "XY and MS Microphone Techniques in Comparison," Sennheiser News Publication, Sennheiser Electronics Corporation, PO Box 987, Old Lyme, CT, USA 06371 (1989).

T. Holman, "The Number of Audio Channels," parts 1 and 2, *Surround Professional Magazine* 2, nos. 7 and 8 (1999).

T. Holman, *5.1 Channel Surround Up and Running*. Woburn, MA: Focal Press, 1999.

U. Horbach, "Practical Implementation of Data-Based Wave Field Reconstruction," 108th AES Convention (Paris, 2000).

D. Huber and R. Rundstein, *Modern Recording Techniques*, 4th edn. Boston: Focal Press, 1997.

ITU (International Telecommunication Union) document ITU-R BS.775.

J. Jecklin, "A Different Way to Record Classical Music," *Journal of the Audio Engineering Society* 29, no. 5 (May 1981).

J. Klepko, "Five-Channel Microphone Array with Binaural Head for Multichannel Reproduction," presented at the 103rd AES Convention, 1997. Preprint no. 4541.

W. Kuhl, *Acustica* 4 (1954), p. 618.

C. Kyriakakis and C. Lin, "A Fuzzy Cerebellar Model Approach for Synthesizing Multichannel Recording, 113th AES Convention (Los Angeles, 5–8 October 2002). Preprint no. 5675.

C. Kyriakakis and A. Mouchtaris, "Virtual Microphones for Multichannel Audio Applications," ICME 2000, New York (October 2000).

E. Madsen, "The Application of Velocity Microphones to Stereophonic Recording," *Journal of the Audio Engineering Society* 5, no. 2 (April 1957).

D. Malham, "Homogeneous and Non-Homogeneous Surround Sound Systems," AES UK *Second Century of Audio* Conference, London (7–8 June 1999).

J. Meyer and T. Agnello, "Spherical Microphone Array for Spatial Sound Recording," 115th AES Convention (New York, 10–13 October 2003). Preprint no. 5975.

D. Mitchell, "Tracking for 5.1," *Audio Media Magazine*, October 1999.

C. O'Connell, *The Other Side of the Record*. New York: Knopf, 1941.

L. Olson, "The Stereo-180 Microphone System," *Journal of the Audio Engineering Society* 27, no. 3 (1979).

A. Previn, *Andre Previn's Guide to the Orchestra*. London: Macmillan, 1983.

O. Read and W. Welch, *From Tinfoil to Stereo*. Indianapolis: Sams, 1959.

E. Schwarzkopf, *On and Off the Record: A Memoire of Walter Legge*, New York: Scribners, 1982.

W. Snow, "Basic Principles of Stereophonic Sound," *Journal of the Society of Motion Picture and Television Engineers* 61 (1953), pp. 567–589.

R. Streicher and F. Everest, *The New Stereo Soundbook*. Pasadena, CA: Audio Engineering Associates, 1999.

G. Theile, "Main Microphone Techniques for the 3/2-Stereo-Standard," presented at the Tonmeister Conference, Germany, 1996.

G. Theile, "Multichannel Natural Recording Based on Psychoacoustical Principles," presented at the 108th AES Convention (Paris 2000). Preprint no. 5156.

G. Theile, "On the Naturalness of Two-Channel Stereo Sound," Proceedings of the AES 9th International Conference, Detroit, Michigan (pp. 143–149).

M. Thorne, "Stereo Microphone Techniques," *Studio Sound* 15, no. 7 (1973).

M. Tohyama et al., *The Nature and Technology of Acoustical Space*, Academic Press, London (1995).

Various, *Stereophonic Techniques*, an anthology published by the Audio Engineering Society, New York (1986).

M. Williams, "Microphone Array analysis for Multichannel Sound Recording," presented at the 107th Audio Engineering Society Convention, New York, September 1999. Preprint no. 4997.

M. Williams, "Unified Theory of Microphone Systems for Stereophonic Sound Recording," Audio Engineering Society. Preprint no. 2466 (1987).

W. Woszczyk, "A Microphone Technique Applying to the Principle of Second-Order Gradient Unidirectionality," *Journal of the Audio Engineering Society* 32, no. 7/8 (1984).

T. Yamamoto, "Quadraphonic One-Point Pickup Microphone," *Journal of the Audio Engineering Society* 21, no. 4 (1973).

"Das Mikrofon," Compact Disc, Tacet 17; Nauheimer Strasse 57, D-7000 Stuttgart 50, Germany (1991).

COMMUNICATIONS AND SOUND REINFORCEMENT

AKG Acoustics WMS300 Series Wireless Microphone System, AKG Acoustics, 1449 Donelson Pike, Nashville, TN 37217 (1997).

G. Ballou, ed., *Handbook for Sound Engineers*. Indianapolis: H. Sams, 2002.

B. Beavers and R. Brown, "Third-Order Gradient Microphone for Speech Reception," *Journal of the Audio Engineering Society* 16, no. 2 (1970).

C. P. Boner and C. R. Boner, "Equalization of the Sound System in the Harris County Domed Stadium," *Journal of the Audio Engineering Society* 14, no. 2 (1966).

C. P. Boner and R. E. Boner, "The Gain of a Sound System," *Journal of the Audio Engineering Society* 17, no. 2 (1969).

D. and C. Davis, *Sound System Engineering*, 2nd edn., Boston: Focal Press, 1997.

D. Dugan, "Automatic Microphone Mixing," *Journal of the Audio Engineering Society* 23, no. 6 (1975).

Y. Ishigaki et al., "A Zoom Microphone," Audio Engineering Society preprint no. 1713 (1980).

D. Klepper, "Sound Systems in Reverberant Rooms for Worship," *Journal of the Audio Engineering Society* 18, no. 4 (1970).

H. Tremaine, *The Audio Cyclopedia*, 2nd edn. Indianapolis: H. Sams, 1969. Various, *Sound Reinforcement, Volumes 1 and 2* (ed. D. Klepper); reprinted from the pages of the *Journal of the Audio Engineering Society* (1978 and 1996).

T. Vear, *Selection and Operation of Wireless Microphone Systems*, Shure, Inc., 222 Hartrey Avenue, Evanston, IL 60202 (2003).

S. Woolley, "Echo Cancellation Explained," *Sound & Communications* (January 2000).

INDEX